MW01491100

DIRECT CURRENT DYNAMO DESIGN

"Lost Technology" Series

reprinted by Lindsay Publications

DIRECT CURRENT
DYNAMO DESIGN

American School of Correspondence
Frank W. Gunsaulus, Editor-in-chief

Copyright (C) 1983 by Lindsay Publications, Manteno, IL

Printed in the United States of America

ISBN 0-917914-05-8

Reprinted from "Modern Engineering Practice" originally published and copyrighted in 1902, 1903, and 1905 by American School of Correspondence

1 2 3 4 5 6 7 8 9 0

THEORY OF DYNAMO-ELECTRIC MACHINERY.

THEORY OF THE GENERATOR.

Definition. A dynamo-electric machine is a machine in which mechanical energy is converted into electrical energy, or *vice versa*, by means of electro-magnetic induction.

The term *generator* is commonly applied to a machine which converts mechanical energy into electrical energy.

A *motor* serves to convert electrical energy into mechanical energy. The generator and motor are similar in construction, but the operation of one is the reverse of the other. The motor is separately considered in a following Instruction Paper.

The operation of a dynamo-electric machine depends upon the relation existing between magnetism and the electric current, and the fundamental principles will now be considered.

FUNDAMENTAL PRINCIPLES.

The Magnetic Field. The space surrounding a magnet is subject to its influence, and is called its magnetic field. This influence is exerted in certain definite directions, and its strength and direction are represented by lines of magnetic force. The strength of the field is indicated by the number of magnetic lines per unit area, and is called its density, or magnetic induction.

Fig. 1 represents the field of a permanent magnet, the lines of force passing outside the magnet from the north to the south pole in curved paths. The number of lines per unit of cross-section, and therefore the strength of the magnet, is greatest at the poles.

Magnetic Field of a Conducting Wire. Whenever a current of electricity passes through a conducting wire, the current creates a magnetic field, which surrounds the wire, as shown in Fig. 2. The magnetic lines of this field encircle the wire and have their greatest density near the same, gradually decreasing in effective strength as the distance from the wire increases. These magnetic lines of force represent a form of energy; and, in fact, a considerable

portion of the energy of the current lies in the surrounding magnetic field.

The direction of the magnetic lines about the conductor
always bears a definite relation to that of the current.   When
looking along the wire in the direction of the current, the direc-

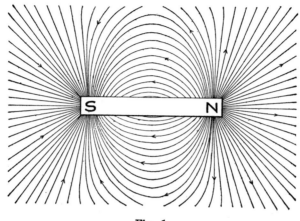

Fig. 1.

tion of the lines of force will always be right-handed; that is,
looking at the left-hand end of the conductor shown in Fig. 2, the
current being from left to right, the lines of force have the same
direction as the hands of a watch.

Field of a Solenoid and Electro-magnet.   When a conductor is made in the form of a coil, and a current passed through

Fig. 2.

the same, the lines of
force, which would other-
wise encircle each turn
of the conductor, unite
and form a strong magnetic field similar to that of a magnet.

The magnetic field of a solenoid is represented by Fig. 3.

When the coil is supplied with a core of soft iron, the
effective strength of the magnetic field is greatly increased.   Such
a combination is called an electro-magnet, and has the properties
of a permanent magnet.

The end at which the lines of force leave the iron core and coil
is called the north pole, and the end at which the lines enter the

iron core and coil is called the south pole. When a person is facing the south pole, the direction of current in the coil is always in the same direction as the hands of a watch; and when facing the north pole, the current is in the opposite direction to the hands of a watch. This gives a very simple rule for determining the

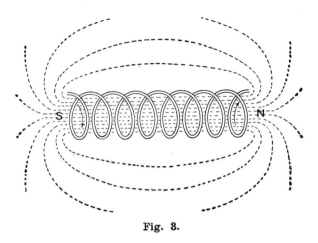

Fig. 3.

relation between the direction of current and polarity of the magnet.

Induced Currents. It would be supposed from the fact that a magnetic field is created by the passage of a current through a conductor, that the reverse action would also take place; that is, the creation of a magnetic field about a conductor would cause the production of a current, and such is found to be the case. The currents created in this manner are called *induced* currents, and are said to be produced by *induction*.

The subject of production of currents by induction is briefly considered in " Elements of Electricity," and the student will note the following facts :

1st. It is only by certain changes of relation between magnetic field and the conductor that the current is produced.

2nd. The change may be due to an increase or decrease in the strength of field about the conductor, and the current flows only while this change in field strength takes place.

3rd. The induced current flows in the opposite direction,

when the field strength is increased to that in which it flows when the field strength is decreased.

4th.   The change in field strength may be produced by the movement of a permanent magnet, solenoid or electro-magnet in the neighborhood of the conductor, — the source of the field is immaterial.

5th.   Relative movement only is necessary, and hence it is immaterial whether the conductor is moved through the field or the field moved across the conductor.

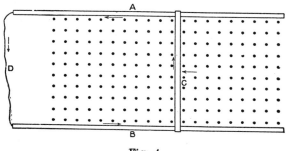

Fig. 4.

6th.   The current flows through a closed circuit only as a result of the induced electro-motive force which is first created, and the strength of the current depends upon the value of the induced electro-motive force and resistance of the circuit, in accordance with Ohm's law.

Movement of Conductor Across the Field.   We shall now consider more fully the production of E. M. F. and current by movement of a conductor *across* a magnetic field.

In Fig. 4 the dots represent the lines of force of a uniform field, their direction being upward through the paper.   *C* represents a conductor which is movable on the rails *A* and *B*, the circuit through *B*, *C* and *A* being completed by the connecting wire *D*.   Now when *C* slides along the rails, it cuts across the lines of force, an E. M. F. is induced, and a current flows, as indicated. The movement of *C* across the field causes a difference of potential between its ends, due to the fact that there is a continual cutting of lines of force about the conductor.   If the slide *C* were moved endwise, however, no E. M. F. would be generated, as no lines would be cut, and during such movement there is no relative change of field with reference to the slide. The same would be true if the slide were moved vertically with reference to the paper. The moment there is any *cutting across* of the lines there is an E. M. F. induced. Such induced E. M. F. exists, however, only during the movement of the slide.

Value of Electromotive Force. The magnitude of the E. M. F. generated depends upon the *rate* at which the conductor cuts across the magnetic lines; that is, it depends upon the number of lines cut in a unit of time. If the conductor cuts across the field in a plane perpendicular to the lines of force, as indicated in Fig. 4, the E. M. F. would depend only upon the field area passed over in a unit of time and upon the density of the field. It therefore follows that the rate at which the lines are cut is determined by the *length* of the conductor, its *speed*, and the *strength* of magnetic field.

Taking, for example, the simple conditions of Fig. 4, let L represent the length of slide C, v its velocity, and B the field density, then

E. M. F. varies as $L \times v \times B$.

The electric and magnetic units are based upon the metric system, and have been so chosen that when a conductor cuts 100,000,000 lines of force per second, there exists a potential difference between its ends of 1 volt. The unit upon which the volt is based is the potential difference generated by moving a conductor 1 centimeter long at the rate of 1 centimeter per second across a field which has a strength of 1 line per square centimeter. This very small unit would be inconvenient for practical purposes, and so the volt was taken as 100,000,000 times this quantity. Hence it follows that

$$E = \frac{L \times v \times B}{100,000,000} = \frac{L \times v \times B}{10^8} = L \times v \times B \times 10^{-8}$$

where E = E. M. F. in volts induced in moving conductor:

L = length of conductor in centimeters;

v = velocity in centimeters per second;

B = number of lines of force per square centimeter.

Therefore, having given the length of conductor, the velocity and the field density, the E. M. F. generated may be computed readily; or if any three of the quantities are given, the fourth may be determined.

Example.— A conductor .6 meter long cuts across a uniform field having a density of 8,000 lines per square centimeter at the rate of 90 meters per minute. What is the voltage generated?

Solution.— Applying the preceding formula, $L = .6 \times 100 = 60$ cm., $v = \dfrac{90 \times 100}{60} = 150$ cm., and $B = 8,000$.

Substituting these values,

$$E = 60 \times 150 \times 8,000 \times 10^{-8} = .72 \text{ volt.}$$

Ans. .72 volt.

Example.— In order to generate a voltage of 1.1 by a conductor 1 meter long, when moving across a uniform field at the rate of 108 meters per minute, what must be the magnetic induction?

Solution.— The known quantities are

$$E = 1.1, L = 1 \times 100 = 100, v = \frac{108 \times 100}{60} = 180.$$

Substituting these values,

$$1.1 = 100 \times 180 \times B \times 10^{-8}$$

Or, $B = \dfrac{1.1 \times 100,000,000.}{18,000}$

$$= 6,111 +$$ Ans. 6,111 + lines.

Direction of Induced Current.   There is a definite relation between the direction of induced E. M. F., or current, the direction of motion and the direction of the lines of force. Referring to Fig. 4, and remembering that the lines of force are assumed to pass upward through the paper, and the slide C to move to the left, the direction of current along C will be that indicated by the arrow. The direction of lines of force, motion and current are all at right angles to each other. If C were moved to the right instead of to the left, the current would flow in the opposite direction to that indicated.

A simple rule for determining the direction of current is the "finger-rule" of Dr. Fleming. This is illustrated in Fig. 5. When the first three fingers of the *right* hand are placed so that they are at right angles to one another, then if the forefinger is pointed in the direction of the lines of force and the thumb in the direction of motion, the middle finger will indicate the direction of induced E. M. F., or current.

For example, if the lines of force were horizontal and passed from left to right and the conductor moved upward in a vertical plane across the lines, the direction of current in the conductor would be away from the observer. Also, if the direction of the lines of force of Fig. 4 were downward through the paper and the slide C were moved to the right, the current would still be in the same direction as that indicated. That is, if the direction of both the field and of motion be reversed, then the direction of current remains unchanged. Reversing *either* the field *or* the motion will reverse the direction of current.

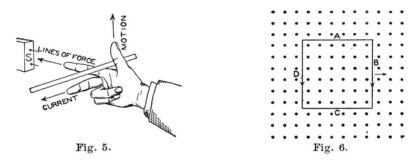

Fig. 5.   Fig. 6.

E. M. F. in Closed Coil. The E. M. F. is generated in the dynamo-electric machine by movement of coils of wire through a magnetic field. We shall now consider the E. M. F. generated in a closed coil when moved in different ways through a magnetic field.

The dots in Fig. 6 represent a uniform magnetic field passing upward through the paper. $A B C D$ represents a closed conducting coil, and let us suppose it is moved to the right, as indicated by the arrow. In such an event, sides A and C cut no lines of force, and no E. M. F. is generated by them. Side B, however, cuts the field, and by applying Fleming's rule it is evident that an E. M. F. is generated in the direction shown by the arrow applied to B. Also, as side D is subject to the same conditions, an E. M. F. is generated in the same direction as that in B, and as each cut the same number of lines of force in the same time, the E. M. F.'s generated are equal. Sides A and C, although not effective in cutting the field, serve as connecting conductors for the other sides. It is clear from the figure that the E. M. F.'s in B and D oppose each other, and therefore no current flows. In a similar manner it may be shown that no current will flow, no matter how the coil is moved, provided it always lies in a plane perpendicular to the magnetic lines. The E. M. F. generated in one part of the coil will always be opposed by an equal and opposite E. M. F. generated in the remaining part.

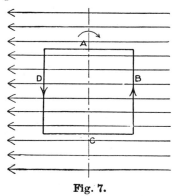

Fig. 7.

When, however, the coil is turned about an axis, the result is different. This may be understood by considering Fig. 7. Here the magnetic lines are represented as passing from right to left, and the coil as being turned about an axis in the direction indicated. The sides A and C cut no lines of force, and therefore generate no E. M. F. For the position shown, side B is moving downward through the paper and cuts directly across the lines and generates an E. M. F. in the direction indicated. Side D is moving upward, and therefore generates an E. M. F. in an opposite direction to that in B. These two E. M. F.'s, therefore, act in the same direction through the coil, and a current flows; the full E. M. F. of the coil being the sum of that generated by each side. This applies to every position of the coil as it rotates about its axis.

While the coil rotates, the number of magnetic lines passing through it is constantly changing, being a maximum when at right angles to the lines, and being zero when parallel to the lines, as in the position shown in Fig. 7. On the other hand, for the movements referred to when considering Fig. 6, the same number of lines passes through the coil at all times, and it may be stated as a general rule, that an E. M. F. is generated in a closed circuit when the same is moved so that the number of magnetic lines passing through it is altered during the motion.

APPLICATION OF PRINCIPLES TO DYNAMO-ELECTRIC MACHINES.

As an E. M. F. is induced in a conductor only while it moves through a magnetic field, thus cutting the lines, it is necessary for the production of a continuous current to have conductors arranged and interconnected so that some are always effectively cutting the field. This result is attained in the generator by rotating such conductors between the north and south poles of a magnet. We shall first consider a single conductor so rotated.

E. M. F. Induced by Rotating Single Conductor. In Fig. 8 suppose a conductor perpendicular to the paper to revolve with a uniform motion on the dotted circle between the $N\ S$ poles, the field having a uniform density. The right-hand portion of the figure will then represent graphically the varying values of the induced E. M. F., and is called the curve of induced E. M. F.

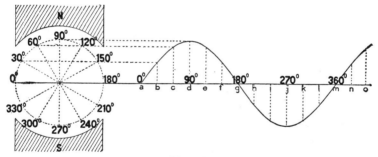

Fig. 8.

The horizontal line is divided into equal parts at points a, b, c, etc., and represent intervals of 30° in the rotation of the conductor. That is, the distance to the curve above or below the line at any point a, b, c, etc., represents the value of induced E. M. F. for the corresponding position of the conductor. For example, when the conductor is at the position 0°, it is then moving *parallel* with the lines of force, and no E. M. F. will be induced. Hence for 0° at a the value is represented as zero, which forms the starting-point for the curve. When the conductor has travelled 90° and is opposite the middle of pole N, it is cutting directly across the field, and at this time generates the maximum E. M. F. The

curve is therefore at its highest point above *d*. At the intermediate points 30° and 60° between 0° and 90° the E. M. F.
gradually increases as represented at *b* and *c*. As the conductor
continues to revolve, a less and less number of lines are cut, until
at 180° it is again moving parallel with the lines, and no E. M. F.
is induced. This condition is represented at *g* where the curve
crosses the line. At intermediate points between 90° and 180°
the E. M. F. has a gradually decreasing value. After passing the
180° position the conductor again begins to cut a gradually
increasing number of lines, but now cuts across them in the
opposite direction to that during the first half of the revolution.
The E. M. F. induced will therefore be in the opposite direction.
Hence at 210° and 240° the E. M. F. gradually increases in an
opposite direction, as shown at *h* and *i*. At 270° the maximum
E. M. F. is being generated in the reversed direction, as represented at *j*. Upon completion of the 360° of one revolution, the
E. M. F. again becomes zero. By continuing the rotation of the
conductor the above series of changes will be repeated during each
revolution. Hence during each revolution the induced E. M. F.
rises from zero to a maximum, decreases, reverses, rises to a maximum in the opposite direction, and decreases to zero again.

Simple Form of Alternating Current Generator. We have
seen that when a coil is rotated in a magnetic field, the opposite
sides aid each other and the E. M. F. is double that produced by
a single conductor. By the use of such an arrangement of a
coil a simple form of alternating current machine is created. Fig.
9 represents the arrangement. The coil is mounted so as to rotate
between the north and south poles of a magnet. The ends of the
coil are connected to copper *collecting rings R R*, mounted on
the shaft with the coil but insulated from each other. Upon each
ring presses a stationary *brush B*, which is always in contact
therewith, and serves to conduct the current to an outside circuit
W. A complete circuit is then made up from one brush through
the external wire *W* to the other brush, then to its collecting ring,
through the coil to the other collecting ring, and to the first brush
again. The alternating current produced by the rotation of the
coil will then pass to the external circuit through the collecting
rings and brushes, and during one-half of a revolution the current

will pass in one direction and during the remaining half in the opposite direction.   The current in the external circuit is available for doing work, and may be passed through motors, lamps, heating appliances, etc.

The Commutator. For the purpose of producing a direct current from the alternating current induced by a rotating coil, a device called a *commutator* is necessary.   Its simplest form is shown in Fig. 10, and consists of a split copper ring, the two *segments* of which are insulated from each other, and each seg-

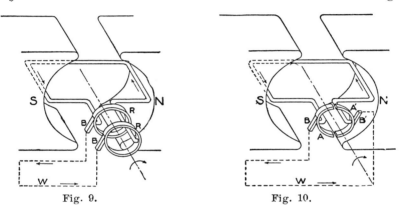

Fig. 9.   Fig. 10.

ment is secured to one end of the rotating coil.   Although an alternating E. M. F. is induced in the coil, by the use of the commutator the current in the external circuit is always in the same direction.   This may be understood from Fig. 10.

For the position of the coil shown, its current will be in the direction indicated and will pass to segment A, brush B, through the external circuit to brush B' and segment A' to the coil.   As the coil continues to rotate, the E. M. F. gradually decreases, until when 90° from the position shown, the E. M. F. is zero.   At this time the segments are located so that the brushes are about to break contact with one segment and make contact with the other.   Further rotation induces an E. M. F. in the opposite direction, but the segments have then passed from one brush to the other, and the direction of current in the external circuit therefore remains unchanged.   That is, when the current in the coil is in the direction indicated, segment A will *deliver* current to brush B, but when in the opposite direction segment A will be under brush B' and *receive* current from it.   The commutator therefore serves to change the direction of the alternating current in the coil to a direct current in the external circuit.

Form of Curve.   The curve of induced E. M. F. in the external circuit when a two-part commutator is used is shown in Fig. 11, and is the same as that in Fig. 8, except that the negative portion of the curve is turned above the horizontal. The direction of the current is always the same, but rises to a maximum and falls to zero twice during each revolution, giving what is called a *pulsating* current.   It must be remembered,

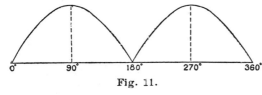

Fig. 11.

however, that the form of curve shown is the theoretical curve, with the assumption that the speed is constant and the field density uniform.   With the actual machine, of course, the conditions are modified more or less, and other considerations, such as self-induction and distortion of the field, materially alter the form of the curve.

The coil, instead of consisting of a single turn, may be made up of two or more, as shown in Fig. 12.   The sum of the electromotive forces induced in each turn is the resultant E. M. F. of the coil.   Hence if there are n turns, the resultant E. M. F. will be n times the E. M. F. induced in one turn.

Owing to the fact that a pulsating current is generally undesirable for practical work a two-part commutator is now seldom used. The current is rendered nearly constant in value, however, by the use of several coils and a commutator made up of a corresponding number of segments.   The coils are interconnected so that the E. M. F. induced in each is added to that of the others.   The various methods of connecting will be considered on following pages, and only the principle will now be considered.

Suppose we have two coils at right angles to each other revolving in a magnetic field, and connected in such manner as to assist each other and yield a direct current to the external circuit. Then, while the maximum E. M. F. is induced in one coil, no

E. M. F. will be induced in the other, and at intermediate points there will be gradual changes. In Fig. 13 the dotted lines indicate the E. M. F.'s induced separately in the two coils, one being 90° in advance of the other. The combining of the two curves produces the curve in full lines, which is formed by taking the sum of the dotted curves. The upper curve therefore represents the fluctuation in E. M. F. of the combined coils, and it is evident that the amount of fluctuation is greatly reduced. In a similar manner it may be shown that a further increase in the number of coils, spaced uniformly, will cause less and less fluctuation, and by using 30 or more coils, a practically constant current may be obtained.

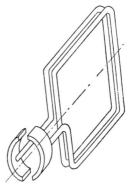

Fig. 12.

Parts of the Generator. Fig. 14 illustrates the essential elements of a generator, which consist of two principal parts,—the stationary *field magnets* and the *armature* which revolves between the poles. A current is passed through the winding around the *cores F F*, to produce the magnetic field in which the

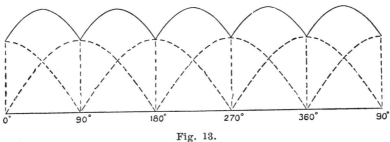

Fig. 13.

coils may rotate. A path for the magnetic lines is provided by the iron cores *F F*, the *yoke Y*, the pole pieces *N S* and the armature as indicated by dotted lines. The armature is made up of the coils and a soft iron core which serves to conduct the magnetic lines. All the coils are interconnected by means of the *commutator C*. The brushes *B B* press lightly upon the commutator and conduct the current to the external circuit, as has been previously explained. The brushes are made adjustable and are carried by *brush-holders* on a *rocker* or *ring* provided for the purpose.

Frequently, and in all large machines, more than two poles are used. Fig. 15 illustrates the magnetic circuit of a dynamo with four poles. Machines having two poles are called bipolar machines, while those with more than two poles are called multipolar. Some machines have as many as 32 poles. The magnetic circuit is always of such form and construction as to produce a strong field in which the armature rotates. For this purpose it is desirable to have the length of the magnetic circuit, or the distance through which the lines of force pass, as short as possible. The material of this circuit is soft iron or soft steel, in order that the field may have great density.

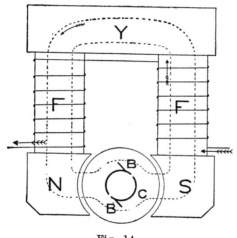

Fig. 14.

Referring to Fig. 14, it will be seen that the magnetic lines pass through the yoke, field cores and pole pieces, then pass across air spaces, called the *air gaps*, and then through the armature core. The armature core is sometimes in the form of a drum and sometimes in the form of a ring, and is always made of very soft sheet iron or soft steel. Since the magnetic lines pass through air with great difficulty compared with the ease of passing through iron, the air gaps are always made as small as mechanical considerations will permit.

The armature is composed of the core above mentioned and the copper coils or windings about the core. The core serves to conduct the magnetic lines from one pole piece to the other; it is mounted upon a supporting frame of cast iron which is firmly secured to the driving shaft and therefore rotates with it. The shaft is usually supported in self-oiling and self-aligning bearings, and must be constructed so that the armature will always be centered between the pole pieces.

Energy Required to Drive the Dynamo. Although there is no mechanical resistance to the rotation of the armature except friction, the mechanical energy required to drive the armature varies with the amount of electrical energy obtained. When no current is delivered, the armature may be easily rotated; but as soon as a current flows, it reacts upon the magnetic field and opposes the rotation, and the greater the current, the greater the energy necessary to drive the machine. The amount of energy necessary to rotate the armature therefore increases as the load put upon the machine increases.

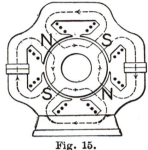

Fig. 15.

ARMATURE WINDINGS.

There are two distinct types of armature windings; one is called the *closed coil* and the other the *open coil* winding.

In the closed coil winding, all the conductors are interconnected so as to form an endless winding, consisting of two or more branches connected in parallel. In the open coil winding the conductors are joined in groups, each group containing in series all conductors which have a similar position with reference to the field. The current is supplied to the external circuit only from those groups which are generating the highest E. M. F. ; all other groups being cut out of the circuit temporarily.

Armature windings may also be divided into *ring* and *drum* windings. In the first, the conductors are wound upon and in the form of a ring; in the second they are wound longitudinally upon a cylinder or drum. Open coil windings are used only in special types of machines, such as those for arc lighting; closed coil windings are used in machines for such common purposes as incandescent lamps, power and heating.

Closed Coil Ring Windings. The ring winding is the simplest, and therefore will be first considered.

Figs. 16 and 17 represent a four-coil Gramme or ring winding having a four-part commutator, and consisting of four coils wound about the iron core, with the junction of the coils connected to the commutator bars. The magnetic lines from the north pole pass to the iron core and then to the south pole, as indicated by the dotted lines, there being no magnetic lines across the space enclosed by the ring. Hence as the armature rotates, the conductors on the exterior surface of the ring cut across the magnetic lines twice during each revolution, while the conductors within the ring merely serve to connect the outside conductors in

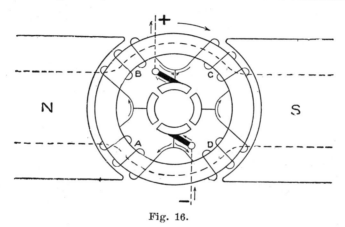

Fig. 16.

series, and are not effective in generating E. M. F. Each coil in Figs. 16 and 17 is made up of three external conductors, and therefore the E. M. F. of each coil will be three times that of a single conductor.

In Fig. 16 coils *A* and *C* are shown as having partly entered the magnetic field, and coils *B* and *D* about to pass out of the field. The path of the current is indicated by the arrows and passes from the external circuit to the lower brush and one bar of the commutator; it then divides and passes through coils *A* and *B* in series on the left, and coils *D* and *C* in series on the right; it then unites, passing to the upper commutator bar and upper brush to the external circuit. As there are two coils in series, the total E. M. F. will be the sum of the E. M. F.'s of the two coils for this position.

Fig. 17 shows the position after the armature has advanced 45° from that shown in Fig. 16. Coils A and C are then generating the maximum E. M. F., and the current passes from the lower brush through coils A and C in parallel to the upper brush. Coils B and D are in what is called the "neutral position," and generate no E. M. F.; in fact, they are temporarily *cut out* of the

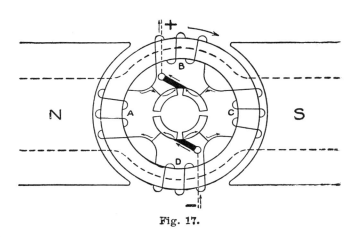

Fig. 17.

circuit, the commutator bars and brushes forming direct connections between coils A and C. Coils B and D are said to be *short-circuited;* that is, there is a direct connection of low resistance between the ends of these coils by the commutator bars and brushes, and no current will then pass through the coils.

Upon further rotation, coils B and D will pass to the other side of the brushes and they

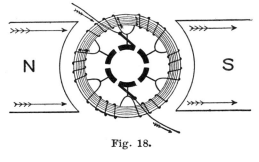

Fig. 18.

will then generate an E. M. F. in the opposite direction.

Fig. 18 shows a ring-wound armature with six coils and as shown, the current divides, passing through the two halves of the winding in parallel. All the coils on one side are in series, and

each add their portion to the total E. M. F. ; the greater the
number of coils, the less will be the fluctuation in the current, as
explained on page 15.   The E. M. F. of each coil varies at different positions, and the E. M. F. generated by some of the coils
is increasing, while that of others is decreasing.   The total
E. M. F., however, remains practically constant when thirty or
more coils are used.

Fig. 19 shows a ring winding for a machine with four poles.
The winding is the same in form as for a bipolar machine, but the

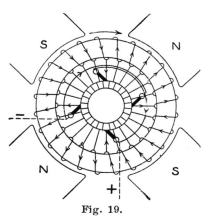

Fig. 19.

number of brushes and number
of circuits through the armature
is doubled.   Therefore there are
two positive brushes and two
negative brushes.   The two positive brushes are electrically
united and connect with the
positive wire of the external
circuit, while the two negative
brushes are united and connect
with the negative wire of the
external circuit.   The current
divides at each of the negative
b r u s h e s, passing in parallel
through four circuits (two for each brush) to the positive
brushes.

For machines with a larger number of poles, there would be
a corresponding increase in the number of circuits through the
armature.

The number of pairs of brushes may also be correspondingly
increased or there may be but one pair since all terminal sections
of the same sign (+ or —) can be soldered together.

Closed Coil Drum Windings.   Drum windings are distin-
guished from ring windings in that all wires of the drum armature
are distributed over the outside surface of the core.   The winding
forms a cage which envelops a cylindrical iron core.   In the ring
armature a large portion of the winding merely serves as connecting leads for that part which lies on the exterior of the core.
In the drum armature, however, the only conductors serving as

connectors are at the ends of the cylinder, and therefore a very different form of winding is necessary.

In the ring winding, adjacent conductors are connected directly in series with each other by looping around the ring core, but in the drum winding the return portion of the loop must lie on the exterior of the core, and in order that the E. M. F. induced in each part shall aid the other, the return portion must lie under a pole of opposite polarity to that under which the forward portion lies, as in Figs. 9 and 10.

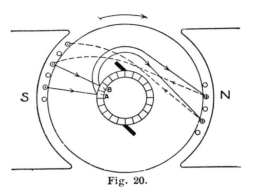

Fig. 20.

The current in passing through the armature from brush to brush must gradually progress through each turn in series. Fig. 20 illustrates a portion of a drum winding, only two complete turns being shown. The small circles represent end views of conductors on the surface of the armature. The crosses and dots indicate the direction of current, the cross representing the tail of a retreating arrow and the dot the head of an approaching arrow. The current gradually progresses from bar to bar of the commutator as indicated by the arrows. The current passes to bar A from a conductor at the left of the core, under the influence of the south pole, then by a spiral connector to a conductor under the north pole, thence by a spiral connector at the back end of the core to a conductor under the south pole to bar B, then by a spiral connector to a conductor under the north pole, etc. It will be noticed that only the alternate conductors are connected in series, but the intermediate conductors are connected in series with each other to the right-hand half of the commutator.

In order to make all end connectors of the same length and form, the commutator is usually turned 90° in advance of the position shown in Fig. 20, giving an appearance as in Fig. 21. This causes brushes of drum-wound machines to be opposite the poles instead of between them as in ring-wound machines.

Fig. 22 gives an end view of a complete drum winding in which there are ten complete turns and ten commutator bars.

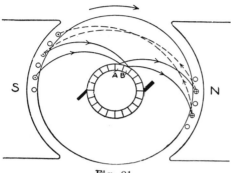

Fig. 21.

The windings form two circuits in parallel from brush to brush, as may be seen by tracing out the paths of the current. All parts of the winding are similar to each other, and a portion, called an *ele- ment*, is made in heavy lines to show clearly the general shape of the turns. Fig. 23 shows a form of drum winding for a four-pole machine with fifteen turns and fifteen commutator bars. It will be seen from the figure that there are four separate circuits between the brushes, two being in parallel from each brush to the other.

Open Coil Wind- ings. Open coil windings are similar to closed coil, except that there are no end connectors between the commutator bars; each coil being alter- nately cut in and then out of the circuit. Such a winding of four coils with four

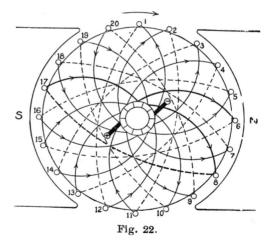

Fig. 22.

commutator bars is shown in Fig. 24. The opposite coils are connected in series and each pair of coils alternately make connec- tion with the brushes.

In General. Only the simplest forms of winding of the

different classes have been considered on the preceding pages, and these were illustrated with figures showing a small number of turns.   In the actual machine, however, the number of turns is much larger, and there are also a great variety of windings under each type, many of which are equivalent electrically but allow considerable variation in the mechanical construction.   The number of

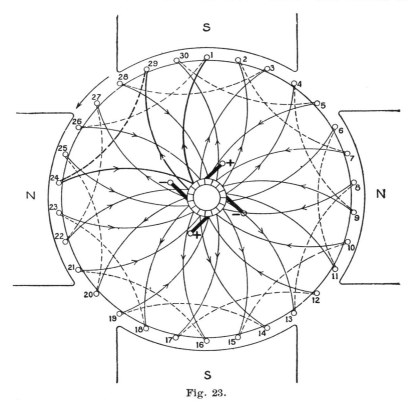

Fig. 23.

poles and the number of brushes used also permit the connecting of the conductors in a variety of ways.

Calculation of E. M. F.   Referring to page 7, it is seen that the value of E. M. F. in volts generated by a conductor moving in a uniform field is given by the formula,

$$E = \frac{L \times v \times B}{10^8} = L \times v \times B \times 10^{-8}$$

in which L is the length of the conductor in centimeters, v the velocity in centimeters per second and B the field strength in lines of force per square centimeter.

From the above formula we can arrive at the voltage of the actual machine. By reference to the figures showing armature windings, it appears that a single armature conductor is opposite a pole, and therefore *active* in cutting lines of force only during certain periods of each revolution. The field opposite a pole is practically of uniform strength, and when a conductor is not

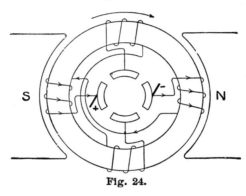

Fig. 24.

under a pole it is not effective in generating any E. M. F. Furthermore, there are usually two or more circuits through the armature connected in parallel, and the voltage of the machine will, of course, be the voltage of each circuit. For bipolar machines there are two circuits connected in parallel, and for multipolar machines there are usually as many circuits in parallel as there are number of poles, although the windings may be varied and interconnected in multipolar machines so that the number of circuits in parallel may be different from the number of poles.

The voltage of any machine will therefore depend upon the number of conductors in series from brush to brush. For a bipolar machine, this will be one-half the total number of conductors on the surface of the armature. Of this one-half only those conductors opposite the pole piece are active. The voltage of such a dynamo will therefore be equal to the E. M. F. generated by a single conductor multiplied by one-half the total number of conductors, multiplied by that percentage opposite the pole pieces. This percentage is usually from 70 to 80. This may be expressed by a formula as follows,

$$E = \frac{L \times v \times B \times \frac{N}{2} \times p}{10^8} = \frac{L \times v \times B \times N \times p}{2 \times 10^8}$$

in which L, v and B equal the quantities previously given, N the number of conductors on the surface of the armature and p equals the percentage of conductors opposite the pole pieces.

Example. A bipolar machine has 800 surface conductors, each 12 centimeters long, and the percentage opposite the pole pieces is 80. The speed is 1,200 revolutions per minute and the diameter of armature is 40 centimeters. The field density opposite the pole pieces is 3,000 lines per square centimeter. What is the voltage of the machine?

Solution. For this case $L = 12$, $v = 3.1416 \times 40 \times \frac{1200}{60} = 3.1416 \times 40 \times 20 = 2513.28$, $B = 3,000$, $N = 800$, and $p = .80$.

Hence,

$$E = \frac{12 \times 2513.28 \times 3000 \times 800 \times .80}{2 \times 10^8} = 289.529 \text{ approx.}$$

Ans. 289.529 volts.

Example. What field density is necessary for a bipolar machine, in order to give a voltage of 220 at 900 revolutions per minute? The armature has 400 surface conductors, is 30 centimeters in diameter and 16 centimeters long, the percentage opposite the pole pieces being 74.

Solution. Here $E = 220$, $L = 16$, $v = 3.1416 \times 30 \times \frac{900}{60} = 1413.72$, $N = 400$, $p = .74$, and B is to be found.

Hence,

$$220 = \frac{16 \times 1413.72 \times B \times 400 \times .74}{2 \times 10^8}$$

or $\quad 16 \times 1413.72 \times 4 \times 74 \, B = 2 \times 10^8 \times 220$

$$B = \frac{2 \times 10^8 \times 220}{16 \times 1413.72 \times 4 \times 74}$$

$$= 6,571 \text{ approx.}$$

Ans. 6,571 lines per sq. cm.

The formula on page 25 may be altered slightly to give the voltage of any machine, depending upon its winding. For example, for a four-pole machine with four circuits in parallel through the armature, the voltage will be that generated by one fourth of the surface conductors, and the formula becomes,

$$E = \frac{L \times v \times B \times N \times p}{4 \times 10^8}$$

It must be remembered that N equals the total number of surface conductors on the armature. For a ring-wound armature N would be equal to the number of turns, and for a drum-wound armature N would equal twice the number of turns.

Factors Determining Voltage. From the preceding it is clear that the voltage of a dynamo depends mainly upon three factors: speed, field strength, and length and number of conductors on the armature. The voltage may be increased by increasing any one of them. The speed of a machine is limited by mechanical considerations, and low speeds are greatly favored for this reason, and also to enable the driving engine and the generator to be directly coupled. The voltage obtained by using a large number of armature conductors is limited by troublesome sparking at the brushes, and the cost is also greatly increased. A high field strength is obtained by having large and powerful field magnets. A strong field greatly increases the weight of the machine and also increases the cost of the iron and field winding, but is advantageous on account of good regulation and reduction of sparking.

It must be remembered, however, that the formulas given above are for the theoretical values only, and these are somewhat reduced by armature reactions which will be considered on the following pages.

THE FIELD.

Magneto=Dynamo. The field of early forms of dynamos consisted of permanent magnets, and such machines were called *magneto-dynamos*. These were objectionable on account of their great weight (a very large magnet being necessary to give even a feeble field) and also on account of the gradual decrease in strength due to vibrations and inability to control the same. Very small machines of this type are still in use for light work, such as ringing telephone bells, but larger machines have been replaced by those having electro-magnets.

Ampere-Turns.   The necessary magnetization of the wrought iron, cast steel and cast iron, which is used for the field frames of dynamos is produced by passing a current through the windings upon the frame, and the strength of the field depends upon the number of turns in the coils and strength of current passing through them.   In fact, the magnetizing force is proportional to the product of the number of turns in a coil and the strength of its current.   This product of number of turns and the current in amperes is called the *ampere-turns*.   As far as the amount of magnetization is concerned, it is immaterial how the number of turns and current strength are related to each other as long as the product is the same.   That is, it makes no difference whether there are 25 turns with 2 amperes, or 10 turns with 5 amperes; the product is 50 ampere-turns.

Saturation of Magnets.   There is a great difference between the relation of the ampere-turns or magnetizing force and the field strength produced thereby for different values in the magnetizing force.   When the number of ampere-turns is small, the magnetization of the iron will increase rapidly for a small increase of magnetizing force, and as the ampere-turns become greater, the resulting field strength increases almost in direct proportion. The increase is not, however, in direct ratio.   This rapid increase in field strength continues to a certain point and then the increase begins to be less and less for an equal increase in ampere-turns, until finally a great increase in ampere-turns produces practically no increase in field strength.   The iron is then said to be *saturated*, or to be in a state of *magnetic saturation*. An increase in the magnetizing force when this point is reached is, of course, of no advantage, and results simply in a waste of energy.

Fig. 25 shows graphically the relation between the magnetizing force and the corresponding field strength for a sample of wrought iron.   In this plot the abscissae or horizontal distances represent the magnetizing force H in ampere-turns, while the ordinates or vertical distances represent the field strength or induction B in number of lines per square centimeter.   The small circles on the curve represent the points obtained by experi-

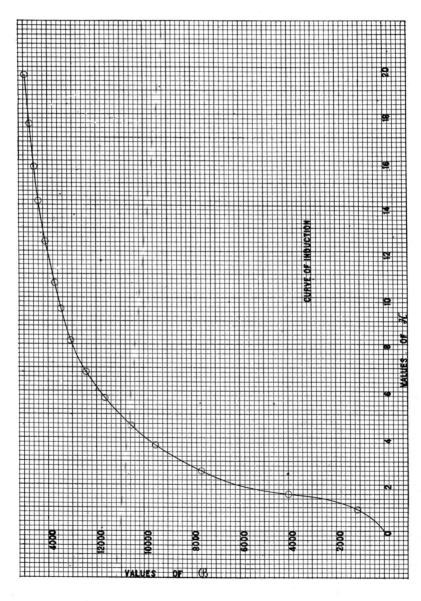

Fig. 25.

ment, and by means of which the curve is obtained. It appears from the curve that an increase in magnetizing force from 0 to .5 does not increase the magnetic induction nearly as much as does the increase from .5 to 1.0, showing that it requires some little force to move the molecules from their irregular position into the more symmetrical arrangement which they are supposed to assume when magnetized. The value of the field strength is about 1,300 lines per square centimeter for 1 ampere-turn, and is 9,900 for 4 ampere-turns, showing the increase to be very rapid. For 8 ampere-turns the field density is 13,000 lines, showing the proportional increase to be much less for the field density than for the magnetizing force, and beyond this point the increase in field

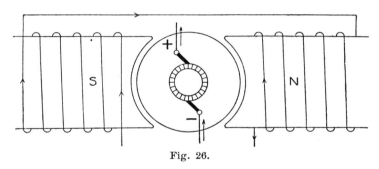

Fig. 26.

strength becomes gradually less and less, until the curve approaches a straight horizontal line, when the iron is saturated.

Types of Field Windings. There are four different types of field windings depending upon the method of excitation, giving rise to the following four classes of dynamos:

1. The *separately excited* dynamo.

2. The *series* dynamo.

3. The *shunt* dynamo.

4. The *compound* dynamo.

Each will now be briefly considered in the order named.

Separately Excited Dynamo. A separately excited dynamo is one having its field coils excited from some outside source, such as a battery or another dynamo. Fig. 26 represents such a machine, the two poles being excited by an outside current, and the current from the armature being available for supplying lamps, motors, etc. In the separately excited machine the field excitation may be maintained nearly constant irrespective of the change in load put upon the dynamo, but it requires an outside current supply. There are but few direct current machines having separate field excitation, but on account of the nature of the current derived from alternating current machines they are usually separately excited. A small direct current dynamo called its *exciter* is used for this purpose.

The Series Dynamo. We now come to consider dynamos which are self-exciting; that is, they supply their own field current and are therefore self-contained. Of these, the simplest is the series dynamo or *series-wound* machine, and in this type the main

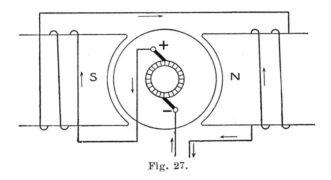

Fig. 27.

current from the armature is passed through the field coils and then to the external circuit. Such a connection is represented in Fig. 27. As these coils carry the main current of the machine they are of comparatively large size, but owing to the large current, only a small number of turns are required to give the necessary magnetization.

The series dynamo has the disadvantage that it cannot be made to start action until a certain speed has been obtained, unless the resistance of the external circuit is below a certain limit. Furthermore, the dynamo requires a special regulator to control the current and voltage with change of load, and about the only practical use for which it is adapted is for operating arc lights.

Shunt Dynamo. In the shunt dynamo or *shunt-wound* machine, the current from the armature has two paths through which to flow. One is through the external circuit and the other is through the field coils. The connections are represented in Fig. 28. The current from the armature divides at the positive brush and passes through the external circuit to the negative brush, and also through the shunt windings to the negative brush,

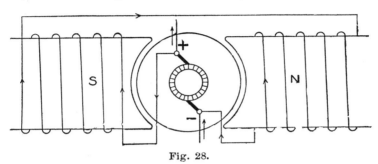

Fig. 28.

the field coils acting as a shunt to the external circuit. In this machine the field windings consist of many turns of fine wire, and take only a small part of the current from the armature; they

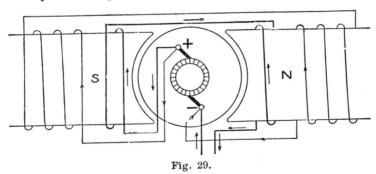

Fig. 29.

differ greatly from the series machine where there are but few turns carrying a large current.

The shunt-wound machine is much better adapted for practical service, such as lighting, and for supplying current to motors, than the series machine, but has been largely replaced by the machine next to be considered.

Compound Dynamo. The compound dynamo or *compound-wound* machine is a combination of the series and shunt windings.

The connections are shown in Fig. 29. In addition to the shunt winding connected from brush to brush, the main current is also passed through a few turns of large wire in series with the external circuit. With this arrangement the disadvantages of the shunt machine are overcome, and a machine which requires almost no attention during changes in load is produced ; that is, the machine is practically self-regulating, and is used almost entirely for supplying current for incandescent lighting and power.

The action of the above considered machines is more fully considered in a following section.

Self-Excitation. As the energy required to excite the field magnets is a very small part of that which the machine is capable of supplying, it is evident that after a machine is once running there is no difficulty in maintaining the field strength. The question naturally arises as to the production of current at start-ing. This is explained by the fact that iron, after it is once magnetized, retains a certain amount of magnetism after the magnetizing force is withdrawn; this is called *residual* magnetism. Hence as the armature is rotated within this weak field, a slight E. M. F. will be generated which is sufficient to supply a feeble current to the field windings. This increases the field strength slightly, which in turn still further increases the E. M. F., and so on until the maximum voltage is attained. This action is called *building up*.

ARMATURE REACTIONS.

On the preceding pages we have considered the fundamental principles upon which the generator is based, and also its general construction. We shall now look into some of the important actions taking place during the operation.

We shall first consider the simple action of the armature unmodified by reactions, and then see how this condition is altered.

Simple Action. Fig. 30 represents an eight-coil ring arma-ture of a bipolar machine. Coils C and G are moving directly across the strongest part of the magnetic field, and are therefore generating the maximum E. M. F. Coils B and F are just enter-ing the field, and D and H are leaving the field and generating less E. M. F. than C and G.

Coils A and E are not cutting across the field, and therefore generating no E. M. F.   Now all coils on the right-hand half of the ring are generating an E. M. F. in the same direction, and therefore assist each other to give the total E. M. F. of the machine, while all coils on the left-hand half are also assisting each other, but generating an E. M. F. in the opposite direction to that of the right-hand half.   The current induced in the outside wires at the right-hand side approaches the observer, while the current in the outside wires at the left recedes from the observer, causing an ascending current in each half of the ring from the lower brush to the upper brush.   It will be noted that each brush for the position shown, bears upon two commutator bars, and the upper brush takes the current directly from coils B and H, while the lower brush conducts the current directly to coils D and F. The thickness of brushes is always sufficient to bridge the insulation be-

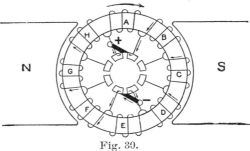

Fig. 30.

tween commutator bars.   Coils A and E are therefore momentarily short circuited or cut out of the armature circuit, and as each coil takes these positions, each is successively short circuited.   The current which the coil formerly carried then seeks the direct path through the commutator bar to the brush.   After being short circuited and carrying no current, the coil is then grouped with the other half of the ring, and a current is set up in the opposite direction through the coil.   Hence as each coil passes from one half of the ring to the other, the current through it in one direction is interrupted, which renders the coil temporarily inactive, and as it passes to the other side of the ring, a current is set up in the opposite direction.   In this manner the commutation of currents takes place.

A plane through the inactive position of the coils is called the *neutral* plane, and coils A and E are said to be in the neutral positions.

Magnetic Field Due to Field Winding. The magnetic field, created by the field coils alone, and unaltered by armature reactions, is represented by Fig. 31. The small circles represent the cross section of each wire, of which only a few are shown, and the direction of current in the field coils is indicated by the dots

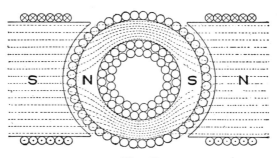

Fig. 31.

and crosses within the circles. In those wires having crosses, the current is flowing from the observer, the cross indicating the tail of a retreating arrow, and in those wires having dots the current is flowing towards the observer, the dot indicating the point of an advancing arrow. The field, created by a current in the field coils only, is uniform and passes through the armature core as shown.

Magnetic Field Due to Armature Winding. When a generator is supplying current, the armature is magnetized by its own current, and the magnetic field in the armature due to its current alone is represented by Fig. 32. The current in each half of the ring magnetizes its half independently of the other, and for the arrangement shown in Fig. 30 each produces a north pole at the lower portion of the ring and a south pole at the top. The two like poles thus created adjacent to each other have a repellant effect and create what is termed a *consequent* pole, causing the magnetic lines to leave the ring at these points, part of the lines passing directly across the ring and part

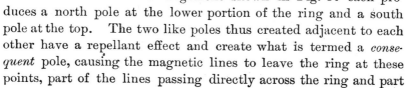

Fig. 32.

passing outside the ring in curved lines from the lower portion to
the top.

When a ring thus magnetized is between the field poles,
the poles will serve as a path for many of the lines, and a field
such as represented by Fig. 33 results.

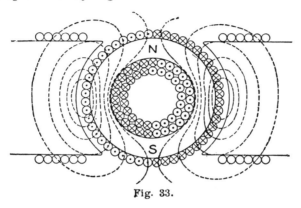

Fig. 33.

Resultant Field.   When a machine is in operation, the magnetism due to both the field and armature windings is present,
and the field obtained will therefore be the resultant of the two
fields represented by Figs. 31 and 33.   These fields, when united,

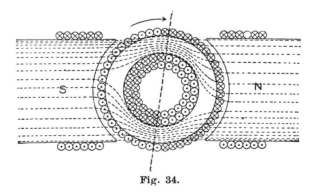

Fig. 34.

will give a resultant field, as represented by Fig. 34.   The field
due to the field windings being the stronger, will still maintain
the same general course, but the effect of the armature is to produce a resultant field in an oblique direction.   It also causes a

crowding of the magnetic lines at the tips of the poles where the
conductors pass from under the same, and a weakening of the
lines at the tips which the conductors are approaching.   It thus
appears that the neutral plane is turned somewhat in the direction
of rotation as indicated by the dotted line.   After passing from
under pole pieces, the conductors generate but little E. M. F.,
and when they are in the neutral plane they generate none
whatever ; therefore from the above it appears that the brushes
should be set in such a position that coils will be short circuited
when in the neutral plane.   Hence the brushes, instead of being
placed midway between the poles in a ring-round machine will be

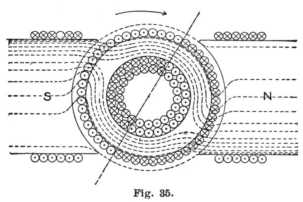

Fig. 35.

shifted slightly in the direction of rotation of the armature.   The
amount which the brushes are shifted from their mid-position is
called the *lead* or *angle of lead.*

This shifting of the brushes causes a still further distortion
of the resultant field, as the field due to the armature winding
alone, instead of being vertical as represented in Figs. 32 and 33,
will be inclined, and coincide with the neutral plane.   This incli-
nation aggravates the distortion, causing the neutral plane to be
shifted to a greater degree than represented by Fig. 34.   The
actual result obtained will therefore be better represented by Fig
35, all conductors on one side of the neutral plane carrying cur
rents in an opposite direction to those on the other side.

There is another cause for additional shifting of the brushes
which causes a slightly greater distortion.   This is due to self-induc-

tion of the armature coils. From the preceding pages it appears that each coil is short-circuited as it passes by the brushes ; the current which the coil carried as it approached a brush ceases to flow, and as it passes to the other side, a current in the opposite direction is set up. On account of self-induction, however, the current which the coil formerly carried tends to continue, and the current in the opposite direction is choked back for a slight interval of time, as self-induction tends to oppose a decrease or increase of current strength. In order to counteract this effect, the brushes are shifted a slight amount in the direction of rotation, so that when each coil is short-circuited it is at that time already somewhat effective in cutting lines of force, so that a slight E. M. F. is generated which hastens the time of decrease in current in one direction, and assists the increase of current in the opposite direction.

We may sum up the effect of armature reaction as follows :

1. The magnetic field due to the armature windings is at right angles to that of the field windings, giving a resultant distorted field.

2. On account of this distortion, and in order to prevent sparking, the brushes must be shifted in the direction of rotation of the armature, and this new position of the brushes in itself produces a still greater distortion.

3. The further shifting of the brushes to overcome the effects of self-induction causes additional distortion.

Shifting of Brushes with Variable Load. When the armature supplies no current and therefore has no load, the field is substantially uniform, as no distortion due to the armature exists. The brushes may therefore be set at the mid-position between the poles. When the load is light, the small armature current causes but little distortion, and therefore the brushes need only a slight lead to secure sparkless running. As the load increases and the armature current becomes greater, the distortion is correspondingly increased, and consequently the proper position for the brushes changes with the load. Hence in the operation of generators, a variation in load will cause sparking at the brushes unless they are shifted to the proper positions. Many machines are now so well designed that but little shifting is necessary from no load to full load, and in some machines the necessity is entirely overcome by special compensating arrangements. Of course, if the field due to the field windings is made very powerful compared with that due to the armature, the distortion will be proportionately less, and in many machines now made, where this plan is carried out, there is little or no need of shifting the brushes with change in load.

Distortion Due to Drum-Wound Armatures. In the above discussion of armature reactions we have referred to distortion due to ring-wound machines only. The same effect in a lesser degree takes place with drum-wound armatures.

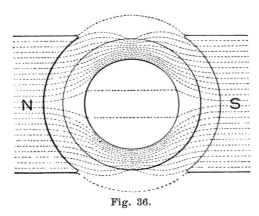

Fig. 36.

Magnetic Leakage. If the magnetic circuit is well designed and made of ample cross-section, practically all the lines of force will pass through the armature core from pole to pole. All lines which do not pass through the core are of course useless, and the E. M. F. generated is less in a corresponding degree. Fig. 36 illustrates the manner in which lines may pass from pole to pole without being of any service. This is called magnetic leakage.

LOSSES IN THE GENERATOR.

There is always some loss in converting the mechanical energy necessary to drive a generator into electrical energy. That is, if 15 horse-power is necessary to drive a certain machine, the equivalent electrical energy derived from the same is always somewhat less. For example, the electrical equivalent of 15 horse-power is $15 \times 746 = 11,190$ watts, but instead of having this amount available at the dynamo terminals, there may be perhaps 90 per cent of this value, or 10,071 watts.

We shall now consider separately the various losses, all of which are dissipated in the form of heat.

These may be classified in two groups, known as the *electrical* losses and the *stray power* losses, and are as follows :

Electrical losses.

    (*a*) Armature conductor $I^2 R$ loss.

    (*b*) Field wire $I^2 R$ loss.

Stray power losses.

    (*c*) Friction of bearings.

    (*d*) Friction of brushes.

    (*e*) Friction due to air resistance.

    (*f*) Waste currents in coils during commutation.

    (*g*) Hysteresis in armature core.

    (*h*) Eddy or Foucault currents in armature core.

    (*i*) Eddy or Foucault currents in pole pieces.

    (*j*) Eddy or Foucault currents in armature conductors.

Of these the electrical losses vary with the load put upon the machine, whereas the stray power losses remain practically constant for all loads. As the stray power losses are always present to the same degree whatever the load may be, it is desirable that they should be as low as possible, since on light loads they may be a considerable proportion of the total power.

Electrical Losses. The electrical losses are those due merely to the resistance of conductors through which a current flows, and may be computed by taking the square of the current passing multiplied by the resistance, or $I^2 R$; where I is the current and R the resistance of the circuit. All circuits carrying a current, whether windings of a generator or mains which distribute power, are subject to this loss of energy. Hence if the resistance and current are known, the power lost may be readily obtained.

For example, if the resistance of a circuit is 80 ohms and the current 40 amperes, the power lost will be $40^2 \times 80 = 128,000$ watts, or 128 K. W. The power lost may also be computed if the current and E. M. F. applied to the circuit are known, by taking the product of the two quantities. This is evident, as from Ohm's law $I = \dfrac{E}{R}$, and substituting this quantity for one I in $I^2 R$, we have $I \times \dfrac{E}{R} \times R = IE$, where E is the E. M. F., applied to the circuit. In the above problem the E. M. F. applied to the circuit is, by Ohm's law, $E = RI = 80 \times 40 = 3,200$ volts, and the loss in watts is, therefore, $40 \times 3,200 = 128,000$ **watts** $=$ 128 K. W., the same as above.

The data for computing the loss in the field windings of a machine can be easily obtained. It is merely necessary to measure the resistance of the windings and the current in the same, or to measure the current and the voltage applied to the terminals of the field windings. The resistance of shunt windings will always be high and the current small, while the resistance of series windings will always be low and the current large.

The data for calculating the loss in the armature is obtained with more difficulty. As the armature is itself generating an E. M. F., the fall of potential in the armature due to its resistance or the *drop* in the armature cannot be measured, and we are therefore compelled to compute the loss from resistance measurements and must use the formula $I^2 R$. The current can be easily measured, but there is some difficulty in measuring the resistance accurately. In this resistance, the armature winding, commutator, brushes and resistance of contacts between commutator and brushes are included. The total resistance is very small, and therefore a slight error in its measurement will make a considerable percentage difference. Also the resistance must be measured when the armature is not rotating, and the measurement obtained may be somewhat different from the resistance when the machine is in operation.

As the resistance of the conductors is greater when hot than when cold, the resistance measurements for calculating the loss in the armature as well as in the fields must be made after the machine has been in operation at full load for a few hours, when all parts will be thoroughly warmed up.

Example.— The resistance when hot of the shunt winding of a certain 120-volt 200-ampere shunt machine is 98 ohms. The resistance of the armature when hot is .019 ohm. What are the electrical losses at full load?

Solution.— The voltage of the machine which is applied to the shunt terminals is 120 volts, and by Ohm's law the current in the shunt windings is

$$I = \frac{E}{R} = \frac{120}{98} = 1.22 \text{ amperes.}$$

The loss in the field winding is therefore $I^2 R = 1.22^2 \times 98 = 146$ watts approximately.

Both the main current and the shunt current pass through the armature; hence the armature current is 201.22 amperes, and the resistance is .019 ohm; the loss is therefore $(201.22)^2 \times .019 = 769+$ watts.

The electrical losses are consequently $146 + 769+ = 915$ watts.

Ans. 915 watts.

Example.— A 500-volt compound-wound generator is designed to supply 650 amperes. The armature resistance is .009 ohm. The current in the shunt winding is 3 amperes, and the resistance of the series winding is .0011 ohm. Find the total electrical loss of the machine.

Solution.— As the current in the shunt field winding is 3 amperes, and the voltage applied to the terminals is 500, the loss in watts is given directly by their product, or $3 \times 500 = 1,500$ watts.

As the main current of the machine passes through the series winding, the resistance of which is .0011 ohm, the loss therein is $650^2 \times .0011 = 464.75$ watts. Likewise the loss in the armature is $653^2 \times .009 = 3,837.6$ watts.

The total loss due to resistance therefore equals $1,500 + 464.75 + 3,837.6 = 5,802$ watts.

Ans. 5,802 watts.

Example.— A 220-volt compound dynamo delivers 300 amperes. The armature resistance is .014 ohm, the shunt winding takes 2.18 amperes, and the fall of potential in the series coil is .54 volt. What are the electrical losses?

Solution.— The loss in the shunt field is $220 \times 2.18 = 479.6$ watts. The current passing through the armature is $300 + 2.18 = 302.18$ amperes, and the armature loss is therefore

$$(302.18)^2 \times .014 = 1,278.4+ \text{ watts.}$$

The loss in the series coil is the product of the lost voltage (i.e., the drop) in it and the current; that is, $.54 \times 300 = 162$ watts. Hence the total electrical losses are $479.6 + 1,278.4 + 162 = 1,920$ watts.

Ans. 1,920 watts.

Friction Losses. The friction losses include those due to mechanical friction of the bearings and of the brushes upon the commutator and also the friction caused by the resistance of the air acting on the rotating armature.

The friction of the bearings varies with the type of machine, whether belt driven or direct connected to the engine, type of bearing, lubrication, etc.

The loss due to bearing friction was formerly considerable, but the bearings of modern dynamo-electric machines have been designed and constructed with great care with the result that friction from this source has been reduced to a minimum. The bearings are usually made self-oiling by means of metal rings which rest upon the shaft and dip into oil which is contained in a reservoir in the bearing. As the shaft revolves, the rings carry the oil up to the shaft. In this manner almost perfect lubrication is secured. In some cases ball bearings have been used with good results; they reduce friction and lessen the cost of lubrication. Ball bearings are not extensively used on account of some mechanical difficulties which have not yet been entirely overcome.

The brush friction is subject to great variation. It is important that the brushes bear upon the commutator with sufficient pressure to insure good electrical contact. Brush holders are usually arranged so that the pressure of the brushes may be varied to obtain the best results. Although brush friction may sometimes be considerable, it usually amounts to a very small percentage of the total loss.

The resistance of the air upon the armature is usually insignificant except when the armature is arranged to act as a fan for the purpose of ventilation, and even then the loss from this source is so slight that it need hardly be considered.

Hysteresis Loss. As an armature rotates, the magnetization of its core is constantly changing at every point. The magnetic lines in passing through the armature from pole to pole cause varying degrees of magnetization in its different parts, and the rotation of the armature brings every part of it successively through these changes. Hence that part of the armature in the neutral plane is magnetized to the highest degree with lines passing through it in one direction; when opposite a pole, the magnetization is a minimum, and when in the next neutral plane it is again magnetized to the highest degree but with lines passing in the opposite direction.   All parts of the armature core are therefore being constantly magnetized in one direction, then demagnetized and magnetized in the opposite direction, again demagnetized, and so on as each pole is passed.   This gives rise to what is called hysteresis loss, which will now be considered.

When a piece of iron is subjected to an increasing and decreasing magnetizing force, the magnetization for equal magnetizing forces will not be the same, and it is found that the magnetism lags behind the magnetizing force, or in other words, the metal has a tendency to maintain any magnetic state which it has once acquired.   This property of iron is called *hysteresis*, from a Greek word meaning "to lag behind."

Fig. 37 shows the hysteresis curve of a piece of wrought iron, or the resultant magnetization when the magnetizing force is varied.   The abscissæ represent the magnetizing force in ampere-turns per centimeter of length, and the ordinates indicate the resultant magnetization in lines per square centimeter.   Abscissæ to the right of the zero point represent positive values and those to the left negative, while ordinates above the zero point are positive and those below are negative.   The curve which begins at *A* shows that the iron had some residual magnetism before any magnetizing force was applied.   As the magnetizing force is increased positively, the magnetic induction gradually increases as in Fig. 37, until point *B* of the curve is reached.   Then the magnetizing force is gradually withdrawn, but the magnetic induction does not fall as fast as it previously rose, for when the magnetizing force is reduced to 10, the induction is nearly 14,000 lines instead of 12,800, as it was on the ascending curve. When the magnetizing force is reduced to zero, the induction is still high, as represented by point *C* of the curve.   Increasing the magnetizing force still further in a negative direction gives an induction curve to point *E*.

A gradual decrease of the magnetizing force to zero and an increase in a positive direction to a maximum gives the curve through points *E F G H*.   It may be noted that point *H* does not coincide with point *B*, although the magnetizing force is increased

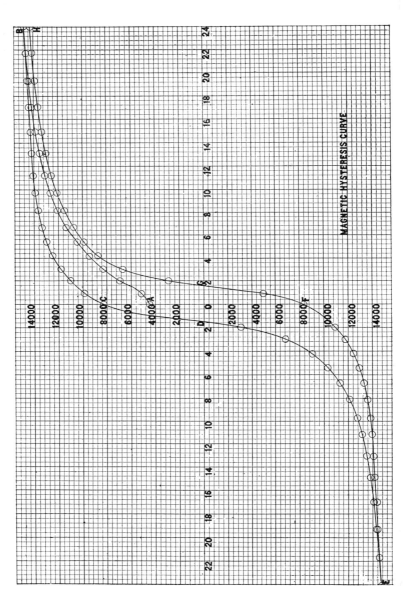

Fig. 37.

from zero to a maximum for both curves. However, in the first magnetization from A to B the induction had a considerable positive value at the beginning, whereas in the curve $F\,G\,H$ it had a large negative value which had to be overcome. If the magnetizing force were now decreased, reversed, increased, etc., a second curve similar to $B\,C\,D\,E\,F\,G\,H$ would be obtained. Such cycles of magnetism as described above always give curves with an enclosed area.

This series of changes is what takes place in the core of an armature, and it is evident that the resultant magnetism always lags behind the magnetizing force as if there were a certain molecular friction of the iron which had to be overcome.

There is, therefore, a certain loss of energy due to hysteresis which is dissipated in the form of heat. The amount of this energy is proportional to the area enclosed by the curve, and it is therefore an easy matter to judge of the comparative magnetic qualities of two samples of iron by plotting the hysteresis curve of each.

The harder the iron, the greater the hysteresis loss. This loss varies as the 1.6 power of the magnetic induction and is also proportional to the rate at which the reversals of magnetism take place, and of course, the greater the amount of metal in which the reversals occur, the greater the total hysteresis loss.

The heating which occurs in the core of an armature due to hysteresis should not be confused with the additional heating due to eddy or Foucault currents. Hysteresis causes heating even if the core is carefully laminated.

Fig. 38 shows two hysteresis curves drawn one over the other, one of which is from a sample of wrought iron and the other from a sample of steel.

The curve enclosing the smaller area is for wrought iron, and the larger for steel. Since the area enclosed by the hysteresis curve is the measure of energy dissipated in heat due to hysteresis, it is evident that the hysteresis loss for wrought iron is much less than for steel. Wrought iron would, therefore, be better adapted for use in armature cores.

The loss due to hysteresis cannot be accurately determined. For very soft iron where the hysteresis curve encloses a small

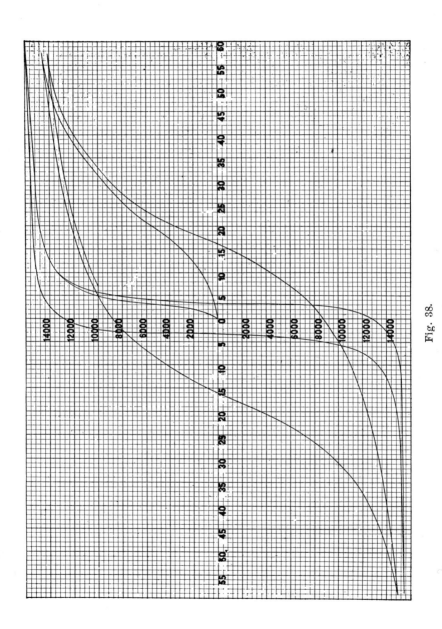

Fig. 38.

area the loss of energy during a complete reversal of magnetism is slight; with hard iron the amount of energy wasted per cycle is somewhat greater, but in practice the loss from this source is usually small.   In order that this loss may be as small as possible the softest wrought iron or steel should be selected for armature cores.

Eddy Current Loss.   We know that whenever conductors are moved in a magnetic field in such a manner as to cut across the lines, or when the field about the same is varied, an E. M. F. is induced which will cause a current to flow if a closed circuit exists.   This principle applies to large metallic masses as well as to the conductors, and therefore currents may be set up in a generator where they are useless, thus causing undesirable heating of parts and loss of energy.   Such currents are called eddy or Foucault currents.

The armature core, if made of one solid mass, is subject to these wasteful currents to an excessive degree, since it consists of a conducting mass revolving in a magnetic field.   It is evident that the outer surface of the armature core cuts across the magnetic field the same as the conductors upon it, and an E. M. F. is therefore induced in the iron itself, tending to set up currents near its surface.   The direction of these currents will be at right angles to the lines of force or the induced current in the exterior of the armature core will be parallel to the current in the armature conductors.   The interior portion of the core serves to complete the circuit for these currents, and therefore innumerable currents, due to this cause, will exist in the core itself.   The E. M. F. induced may be slight, but the resistance of the core is so small that the resultant currents may be large.   These currents are always strongest near the exterior surface of the armature core. This eddy current loss described above varies as the square of the speed and also as the square of the magnetic induction.

Lamination.   The value of such currents can be decreased by reducing their E. M. F. or by increasing the resistance which is offered to their flow.   These results can be accomplished by constructing the armature core of thin sheets of iron carefully insulated from each other.   These thin sheets or discs are placed perpendicular to the axis of the armature, thus being perpendicular to the direction of induced currents, but parallel with the lines of force passing through the core. Consequently they do not interfere with the passage of magnetism from pole to pole, nor increase the magnetic resistance of the armature. They do, however, effectually interrupt and reduce the eddy currents because such currents can then exist only within each separate disc, as shown in Fig. 39, and if the discs are made sufficiently thin and the insulation between them is good, the loss due to wasteful currents is slight. An armature built up of discs or thin sheets of iron is said to be laminated. The thickness of the discs should not exceed two millimeters, and they should be insulated by varnished paper, enamel, or by being slightly oxidized on the surface.

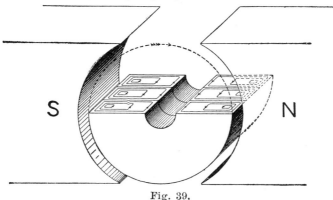

Fig. 39.

Eddy Currents in Pole Pieces and Armature Conductors. Polar tips in which magnetization varies become heated. Hence tips are now laminated, the laminations being parallel to the lines of force but perpendicular to the direction in which currents would be induced. Armature conductors are also subdivided for the same reason, and the loss in eddy currents is reduced to a total of usually less than one-half of 1%.

DIRECT CURRENT DYNAMOS.

Having become familiar with some of the fundamental principles and laws governing the operation of a dynamo by a study of electro-magnetic induction, one is prepared for a more detailed study of the machine. The three important types of dynamos, the series, the shunt and the compound, have already been briefly explained, but will now be considered more in detail, and their properties will be more fully explained and illustrated by diagrams.

All of the important parts of the dynamo, such as the armature, the commutator, the brushes and brush-holders, and the field will now be treated separately and in detail.

In general, dynamos are used to furnish either light, heat or power. A dynamo receives at the pulley a certain amount of mechanical energy, which it transforms into electrical energy and delivers through the circuits with which it is connected. The power in a circuit is equal to the product of the difference of potential, E, between the terminals of the dynamo and the current, I, flowing in the circuit, or $E\,I$. When this product, $E\,I$, is great, a large amount of work is expended in the circuit, and to express this fact it is said that a heavy load has been put upon the dynamo. The value of $E\,I$ is increased by increasing either E or I, or both; that is, the load on the dynamo is increased when a larger current comes from it, or when the same current flows, but at a higher voltage.

We have seen that electrical power is measured in units, to which the name **watt** is given, a watt being the power of 1 ampere under a pressure of 1 volt; the product of the number of amperes and the voltage of a circuit is therefore its power. A dynamo which maintains a difference of potential between its terminals of 10 volts, and which delivers through the circuit connected with its terminals 5 amperes of current, is furnishing energy at the rate of 50 watts. Had this same dynamo been working at a difference of potential or pressure of 25 volts, and had it been delivering only 2 amperes in the circuit, it would have still been working at the rate of 50 watts. It makes no difference in the amount of power what the pressure or the current is, as long as they are of such magnitudes that their product remains the same; the dynamo will then furnish the same amount of energy per unit of time, and consequently it will take the same pull on the belt, and the same effort in the engine to drive it.

These considerations suggest two methods for distributing power and light. The necessary condition of supply is that the dynamo must be ready to meet a demand which varies between certain limits. This may be done by causing the dynamo to keep the voltage at its terminals constant at all times, and to vary the current according to the demand; or the dynamo may keep the current supplied constant, and vary the voltage at its terminals according to the demand for energy. It will make no difference to the driving power which is done. It has, however, for many reasons, been found most convenient to supply incandescent lighting systems by the first method, and arc lighting systems by the second. Power distribution has occasionally been accomplished by the second method, but the first is generally used.

In the study of **Direct Current Dynamos** it will therefore be of advantage to divide the subject into two parts: (1) Constant Potential Dynamos, machines that supply current at a constant pressure for all loads; and (2) Constant Current Dynamos. machines that supply a current of constant strength for all loads.

By having more lamps or running more motors on the circuit we increase EI, or the load upon the dynamo. In constant potential machines, E remains constant for any change in the number of lamps or motors on the circuit, within the limits of the capacity of the machine; hence, to increase the load on the dynamo, I must be increased, and it is seen readily enough that the load is proportional to I if E remains constant, or since I is inversely proportional to R, the resistance of the circuit, the load on a constant potential machine is inversely proportional to the resistance of the circuit.

In a constant current machine I is constant and E must be increased if $E\,I$, the load, is to increase. If E rises, and I remains the same, then R must increase according to Ohm's law. Hence the load on a constant current machine is proportional to E and to R. An increased load on constant current machines requires larger values of E and R; on constant potential machines it requires a larger value of I and a smaller value of R.

If E is to remain constant, the magnetization produced by the field coils must remain constant, or, in other words, the conductors on the armature must cut the same number of magnetic lines of force each second. If I is to remain constant and E is to increase, then a correspondingly greater number of lines of force must be cut each second.

The essential difference between the two types of machines then is that in one there is a nearly constant magnetic flux for all loads, and in the other there is a magnetic flux varying with the load. The two types of constant potential and constant current dynamos will be considered separately.

Characteristic Curves. A characteristic curve, as its name indicates, is one which shows the characteristics or peculiarities of the dynamo to which it belongs. Such a curve is obtained by plotting the corresponding values of volts and amperes of the dynamo at different loads. The data for such a curve are obtained by taking simultaneous readings of current and voltage at various loads while the dynamo is in operation. Upon a piece of cross-section paper the amperes are plotted as abscissæ and the volts as ordinates; that is, the number of amperes of one reading are laid off horizontally from the lower left-hand corner, called the origin, covering as many divisions as there are amperes; and from this point a distance is laid off vertically, covering as many divisions as there are volts corresponding to that number of amperes. The point thus obtained is a point on the curve. The remaining readings are laid off in like manner, obtaining a number of points; a smooth curve through these points gives the desired characteristic.

A number of very important properties of a dynamo may be shown by means of the **characteristic curves.** These same characteristics may generally be figured out mathematically, but it is more correct and more satisfactory to get at the results from a practical test of the finished machine, as a general rule. These characteristic curves explain the actions and possibilities of a dynamo in very much the same way that the indicator card of a steam engine gives its desired information. By drawing the characteristic curves to some known scale, the horse-power at which the dynamo will work with the greatest efficiency can be calculated.

Having drawn the characteristic curve of a machine to some particular scale, and knowing either the current, I, or voltage, E, at which the dynamo is working, the other value may be read off the curve directly. The value of the current, multiplied by the corresponding voltage, gives the output of the machine in watts, and this divided by 746, the number of watts corresponding to 1 horse-power, gives the output of the machine in horse-power.

It is possible to plot horse-power curves on the same paper with the characteristic curve (see Fig. 3), since horse-power $=$ $\dfrac{volts \times amperes}{746}$. Any combination of amperes $\times$ volts which will give 746 is equal to 1 horse-power; thus 74.6 amp. $\times$ 10 volts $=$ 1 horse-power. Evidently, then, there is an infinite number of points the product of whose co-ordinates equals 746, and a line drawn through these points is a curve of 1 horse-power. If the characteristic of the dynamo cuts the curve at any point, it means that at that point the dynamo is furnishing 1 horse-power of electrical energy. It follows at once that curves may be similarly drawn for 2, 3, 4, or any number of horse-power. Then a glance at the characteristic curve will reveal at once what the activity of the dynamo is.

THE SERIES DYNAMO.

After the invention of the magneto machine with its permanent steel magnet, it was not a very great step to put coils of wire on the field magnets, and cause the current, as it comes from the armature, to pass through them before being led to the outside circuit, thus obtaining self-excitation (see Fig. 1).

A machine so connected is called a **series dynamo** because all its conducting parts are in series with one another. The properties

of this machine can be best explained by considering its characteristic curve. Every individual machine will of course have a curve
differing somewhat from those of other machines which are differ-
ently proportioned, but the same general features will be present
in the curve of all series dynamos. There are three characteristic
curves possible for every dynamo, the **external,** the **internal** and
the **total.** The distinction is as follows: the electromotive force
generated by the machine is used up in two ways; in forcing the
current through the resistance of the outside circuit, and in forcing it through the internal resistances of the dynamo itself.
Therefore the voltage measured by a voltmeter applied to the
terminals of the dynamo is the voltage which is forcing the cur-

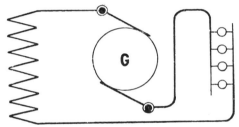

Fig. 1.   Diagram of Series-Wound Dynamo.

rent through the outside circuit. This voltage is given by the
lower curve in Fig. 2, which is called the **external characteristic.**
To get the total electromotive force generated by the machine it
is necessary to multiply the current by the internal resistance
and add the result to the voltage measured at the terminals. The
upper curve in Fig. 2 gives the total E. M. F. and is called the
total characteristic. A characteristic curve is external, internal,
or total according as the volts plotted against the amperes
are the volts measured at the terminals, the volts lost in the
armature, or the total volts. From the series characteristic
curve shown in Fig. 2 it is evident that when the circuit is open
and there is no current, there is no voltage generated by the
machine. With the current equal to zero, the strength of the
field due to the field magnets is zero. Were there any permanent
magnetism remaining in the iron as there generally is, there would
be a slight electromotive force generated, and the curve would

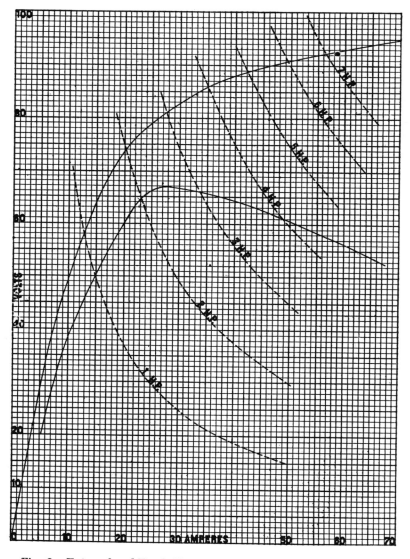

Fig. 2. External and Total Characteristic Curves of Series Dynamos.

not start at zero but a little way up on the scale. A dynamo with no magnetism whatever could not be made to take a load, or *build itself up* as the action is called. The curve runs up very steeply and approximately in a straight line, since the field, and consequently the electromotive force, increases with the increase in the current. This action goes on until the iron approaches saturation, or the point where it cannot take up any more magnetism. As it nears this point the curve bends over more and more, though still rising, until the actual point of saturation is reached. Beyond this point any increase in current cannot increase the strength of the field much and another factor becomes of importance, namely, the armature reaction. The increase in current must of course increase the armature field which tends to weaken the field due to the field magnets and consequently the voltage of the machine. It should be understood that the field of the armature does not actually exist because it is overpowered by the field magnets; it would exist were the field magnets absent and the current maintained by some external means. Its effect is to distort and weaken the field due to the field magnets. Beyond the saturation point therefore, any increase in current cuts down the field and consequently the electromotive force generated in the armature, provided the speed remains the same.

There is a very interesting property of series dynamos, which is best explained in connection with the characteristic curve. There is for every series dynamo a certain value of the current called the *critical current* at which the machine is unstable. Any increase in the circuit resistance tending to decrease the current below this value will result in the complete *unbuilding* of the machine and the consequent loss of load. Similarly the machine will not take a load or *build up* unless the resistance of the circuit is low enough so that the current can reach the critical value.

An examination of the characteristic curve shows that at a point a little more than two thirds of the way up to the maximum value of the curve, any decrease in the current of the machine decreases the voltage much more rapidly than is the case at any point higher up. This large decrease in voltage still further decreases the current, which result reacts on the voltage again, and this process goes on until the machine is completely unbuilt.

Total and external characteristic curves of a small Wood arc light dynamo are shown in Fig. 3. The total voltage generated by the armature is equal to the voltage observed at the terminals plus the lost voltage of the armature and series coils.

This was calculated by the equation

$$E' = E + I\,(r_a + r_{se})$$

where r_a was 5.4 ohms and r_{se} was 6.83 ohms.

The observed voltages are corrected for a speed of 1150 revolutions.

The electrical efficiency for this machine was calculated and found as follows:

DATA FOR EFFICIENCY OF SMALL WOOD DYNAMO.

E	Speed	I	$E'\ I'$	$E\ I$	$\dfrac{E\ I}{E'\ I'}$
88	1150	1	100	88	88 %
298	1150	4.4	1548	1311	84.7%
310	1150	4.6	1684	1426	84.6%
372	1150	6.0	2670	2232	83.5%
402	1150	7.0	3413	2814	82 %

The usual use of the constant current dynamo is for arc lighting. Arc lamps are commonly run in series and require about 10 amperes, and a potential of about 50 volts across the terminals of each. To maintain a constant current in the circuit the E. M. F. at the brushes of the generator must then increase 50 volts as each lamp is thrown in.

Constant current dynamos are series wound. Of such machines, we know that with a variable resistance in the external circuit, the current, or E. M. F. or both will vary. In the case of a machine supplying a certain current, if the resistance is increased the current strength will fall, and the field coils being in series the magnetic field is weakened, thereby lowering the E. M. F. and still further lessening the current. On the other hand if the external resistance is reduced, a larger current flows and a correspondingly higher potential is generated. Obviously, if such a machine is required to give a constant current through a

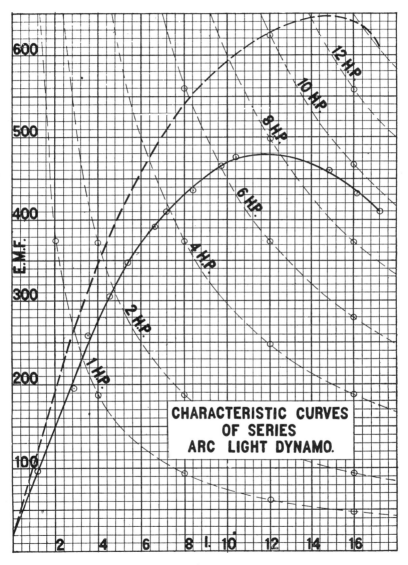

Fig. 3.

variable resistance, there must be some means provided for raising and lowering the E. M. F.

For effecting such regulation there are these three methods: (1) Varying the speed. (2) Varying the field strength. (3) Changing the position of the brushes.

The first method is seldom made use of although in principle it is very simple. There must be an automatic regulation of speed and the E. M. F. of course rises and falls with it. Such regulation, on account of the inertia of the heavy moving parts of the engine and generator, is too slow for lighting service.

Regulation by the second method is obtained either by changing the number of active conductors in the field or by changing the connections of the coils from series to parallel and *vice versa*. Running on light load (few lamps in series) only enough turns in the field are left in circuit to maintain a current of 10 amperes, and as more lamps are added to the circuit more field turns are thrown in to strengthen the field and to raise the voltage. The disadvantage of this method is the small range of loads that can be economically carried. For as the field coils are thrown in the cores soon rapidly approach the point of saturation, beyond which a large increase of ampere-turns effects the E. M. F. very slightly. On the other hand the machine will not run well on very light loads, for the current being constant the reaction of the armature is constant and on the weakened field has a greatly increased effect of distortion upon the field and causes serious sparking at the brushes.

The third method, that of shifting the brushes, is the most common one. To understand the principle of this regulation it will be well to refer back to the figure showing the two paths of the current through a ring armature and out by the brushes (see illustration in section devoted to " Theory of Dynamo-Electric Machinery "). It will be seen readily that the maximum E. M. F. at the brushes is obtainable when they are at the neutral point, that is when all the amature coils on either side of the brushes generate an E. M. F. in the same direction. If the brushes are shifted from the neutral point then some of the coils on either side oppose the others on the same side and a reduced pressure at the brushes is the result. If the brushes were placed at points midway between the neutral points, then on each side half the coils would oppose the other half, and the pressure at the brushes would be nil; that is, the algebraic sum of the E. M. F.'s of the coils in each half would be zero. Then by moving the brushes from this point toward the neutral point, this sum, or the E. M. F. of the machine, would increase gradually to the maximum. Such is the principle of this method of regulation.

A constant current machine running on light load will have its brushes in a position considerably off the neutral point, and as more lights are thrown into the circuit, making the load heavier, they will be brought correspondingly nearer, maintaining sufficient E. M. F. at the brushes to force a current of constant value through the increasing external resistance. The shifting of the brushes is done automatically by electro-magnets in circuit, which act instantly to counteract any change of current strength flowing through them.

In practice various devices are used by different makers to accomplish the regulation. In some machines the position of the brushes on the commutator is altered by a mechanism controlled by the current; in others the controlling mechanism alters the field strength by short-circuiting some of the coils, or by increasing or decreasing a resistance placed as a shunt around the field, thus varying the value of the current in the field coils instead of the number of turns. The variable brush method is sometimes combined with the variable field method. A description of some forms of arc lighting machines, together with their regulators, is given in a following section.

Saturation Curve. To obtain a certain magnetization by a field coil, it is immaterial whether the coil consists of a small number of turns carrying a large current, or of a large number of turns carrying only a small current. As long as the product of the number of turns and amperes passing in them, or **ampere-turns,** is the same, the magnetization produced will be of constant value. Thus a coil of 100 ampere-turns may be made up of one turn of a large heavy wire carrying 100 amperes, or it may be made up of 100 turns of fine wire carrying one ampere, and the magnetizing force which will be exerted will be exactly the same in either case. Magnetizing forces are therefore sufficiently expressed by the number of ampere-turns in the magnetizing coil.

The saturation curve of a dynamo shows to what intensity the field magnets have been magnetized, and thus whether they are properly designed. The curves are formed by plotting points whose abscissæ are values of ampere-turns in the field coils, and whose ordinates are the corresponding values of the voltage, these points being plotted for varying values of the current. The readings may be observed for both ascending and descending values of the current, thus showing the effects of hysteresis.

From an economical standpoint it is best not to carry the magnetic flux too near the point of saturation during normal working, for the number of ampere-turns required to produce a given magnetic flux is greater in a saturated field than in one not saturated. In certain types of machines, however, there are certain advantages in practical points of working and regulation to be derived from having saturated fields that are of such importance as to outweigh the value of greater efficiency. By knowing from the type and the design of a machine to what degree its fields should be saturated, and by having an experimental saturation curve of the machine, a comparison can be made between the ideal and the result. Fig. 4 gives the saturation curves of the magnetic circuit of a Crocker-Wheeler dynamo for no load and with load, and for increasing and decreasing values of the current, the dynamo running at a constant speed of 1,620 revolutions per minute. The total characteristic curve of the simple series dynamo is only a slightly modified saturation curve, as will be seen from a comparison of Fig. 2 and Fig. 4. But in dynamos provided with regulators, the operation of these appliances changes the form of the characteristic very markedly.

THE SHUNT DYNAMO.

The shunt dynamo is so named because the coils which excite the field, instead of being connected in series with the outside circuit, form a shunt about it, the field circuit and outside circuit being connected in parallel at the brushes (see Fig. 5).

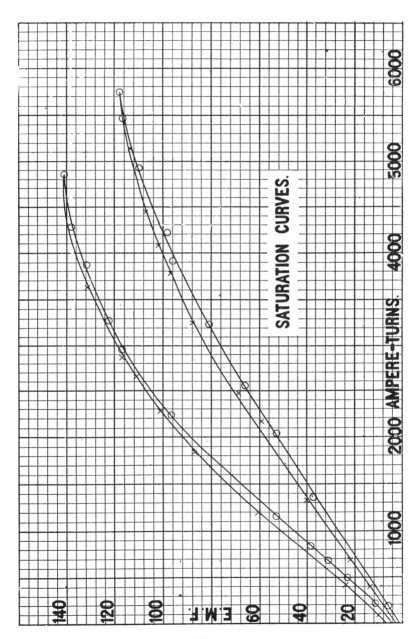

Fig. 4.

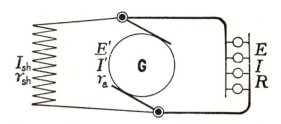

Fig. 5.   Diagram of Shunt-Wound Dynamo.

As in the case of the series dynamo, the properties of the shunt machine can be best explained with the aid of characteristic curves (see Fig. 6).   There are three characteristic curves to be considered, the **external,** in which the volts at the terminals are plotted against the amperes in the outside circuit, the **internal,** in which the volts at the terminals are plotted against the amperes in the shunt circuit, and the **total** in which the total volts generated by the machine are plotted against the total current.   The total characteristic differs from the external only in that the volts are obtained by adding to the volts at the brushes the drop in potential through the armature.   This drop in volts is equal to the current passing in the armature multiplied by the armature resistance.   The total current is obtained by adding the current in the shunt circuit to that in the outside circuit.   The internal characteristic taken with the outside circuit open is like the curve for a series dynamo. It rises steeply at first and then bends over as the saturation of the magnets begins to be noticeable. In fact with the outside circuit open the machine is a series dynamo with all its circuit in the field coils.

The external characteristic curve is a loop.   It begins at maximum voltage and zero current, and extends almost horizontally at first, then slopes down more and more, turns rather suddenly, then doubles back and tends to approach the origin.   When the dynamo is started the outside circuit is open and there is no current through the armature.   As soon as it is built up the armature is generating only the current for the field coils which, as we have seen, is very small owing to the great resistance of these coils. But the small current produces only an insignificant drop in the armature so that the voltage at the brushes is nearly the total generated by the armature.   Hence the current in the field coils is as great as it can ever become which makes the field magnetism and consequently the E. M. F. generated by the armature a maximum.   Now suppose that a number of lamps (connected in

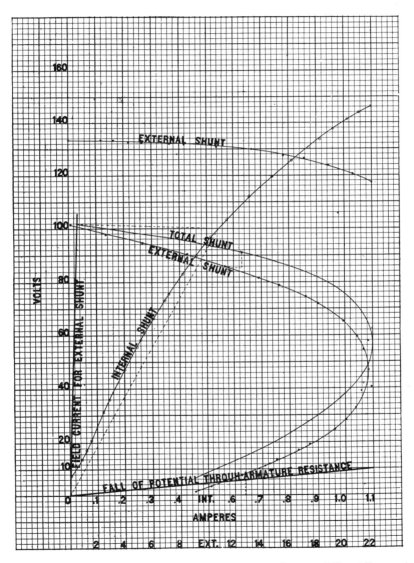

Fig. 6. Total. External and Internal Characteristic Curves of Shunt Dynamo.

parallel across the leads from the machine) are thrown on the machine. A current flows at once, and as a result there is a drop through the armature so that there is a little less voltage available at the brushes to excite the field coils, which means a slight dropping of the total voltage. If more lamps are connected in parallel, the existing voltage at the terminals will act for an instant, and furnish them with as much current as those already on are receiving, but the greater current means increased drop in the armature which, as explained above will again bring down the total voltage generated. With less voltage at the terminals the total current will diminish a little. Therefore lowering the resistance of the circuit by the addition of lamps in parallel lowers the available voltage, because it increases the amount of the volts lost in the armature. By building the armatures with an extremely low resistance it is possible to make machines for which the drop in terminal pressure is not very noticeable over quite a large range of load. It is for this reason that a shunt machine is used for incandescent lighting where it supplies an approximately constant voltage. The dynamo would be designed for such a voltage that at low loads the voltage would be a little too high while at heavy loads it would be a little too low. In actual lighting work, however, the resistance of the field coils is made somewhat less than would·be necessary to give the right voltage. This is accomplished by using a larger size of wire than necessary, the number of turns remaining the same.

By inserting a **rheostat,** or resistance which can be varied at will, between the field coil and one of the brushes, the current in the coil can be cut down to the proper value under normal conditions, and when the load increases and the effective voltage tends to drop, it is only necessary to cut out some of the resistance in the rheostat to obtain more current, and consequently greater excitation in the field coil, thereby raising the voltage to its normal value again. The rheostat (see Fig. 7) consists generally of a resistance with connections at short intervals along it and terminating in metal contacts arranged in a circle. A sliding contact piece is moved over them by the handle, introducing the current at any point on which the sliding contact is resting. In this way the amount of resistance in series with the field coil may be regulated with great exactness.

This regulation is generally effected by hand, the resistance being decreased when the voltmeter shows that the voltage is falling on the line, and vice versa. By means of regulators, magnets, solenoids and various electrical and mechanical means this regulation is sometimes made automatic.

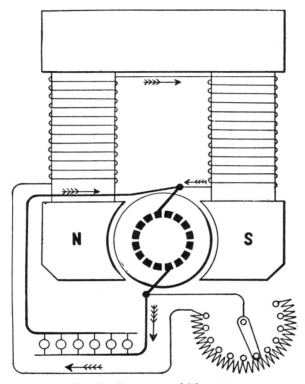

Fig. 7. Dynamo and Rheostat.

Returning to the characteristic, it is interesting to note that the continued increase in load on the machine, due to decrease of external resistance, causes the voltage to fall more and more rapidly until a point is reached where a decrease in circuit resistance can throw no more load on the machine, because the decrease in voltage is so rapid as to unbuild the machine. Theoretically the curve would come down to the origin; it does not, practically, however, owing to residual magnetism.

The resistance of the outside circuit corresponding to any point on the characteristic curve can be found very readily. The voltage divided by the current gives the resistance. By drawing a straight line (see Fig. 8) from the origin or the point where the current and voltage are 0 through the point corresponding to 50 volts and 50 amperes it will intersect the characteristic curve at a certain point. It is obvious from the figure that the voltage at this point just equals the current, and so the resistance here is 1 ohm. Similarly by drawing a line from the origin through the point where voltage is 100 and current 50 the characteristic curve is intersected at another point and the resistance of the circuit at this point is 2 ohms. A scale of resistance may be marked on the vertical line corresponding to 10 amperes. By drawing a line from any point on the characteristic curve to 0, the point of intersection on this resistance scale line gives the resistance of the outside circuit. The characteristic curve shown in Fig. 6 is for a small shunt dynamo which being run at 630 revolutions gave a maximum of a little less than 2 horse-power at 68 volts and 19.2 amperes. With a decrease in resistance much below $2\frac{1}{2}$ ohms the current increases but little, whereas the voltage falls a great deal. If the external resistance becomes less than 1 ohm the machine loses its voltage and current immediately and will not build itself up again until the resistance is increased. The most critical part of this curve is where the voltage is about 30 or 31 and any given change of resistance at this point will alter the voltage more than at any other part of the curve. By increasing the resistance the voltage is steadily increased until it gets to its maximum when the external circuit is open and the resistance is infinite, the whole voltage being then available for magnetizing the shunt field coils to their maximum strength.

Fig. 8 shows the characteristic curves of a shunt-wound Gramme dynamo capable of giving 400 amperes at 120 volts. The armature conductors were not capable of carrying more than 400 amperes, and the part of the curves not actually found by experiment is shown in dotted lines. The lower curve marked E is the external characteristic while the upper curve marked E' is formed from it by adding to the external voltage at any point the corresponding value of the voltage lost in the armature at the given time.

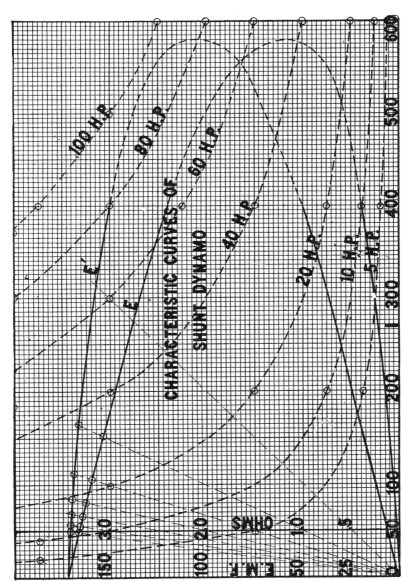

Fig. 8. Characteristic Curves of Shunt-Wound Gramme Dynamo.

Both series and shunt machines have critical points, but the series machine will act only with the resistance of the circuit less than a certain amount while the shunt machine will work only with it greater than a certain amount. With the series machine, increase in the number of lamps in series decreases the ability of the machine to maintain its load, increase in the number of lamps in parallel increases the ability of the machine to generate and consequently increases the current per lamp. With the shunt machine an increase of the lamps in series increases the ability of the machine to furnish current, while increase in the number of lamps in parallel decreases the ability of the machine to supply current. These opposing characteristic qualities are made use of to secure a machine which is automatic in its ability to furnish a constant potential under widely varying loads.

THE COMPOUND DYNAMO.

Electrical energy for incandescent lights, power circuits, and railways is supplied by constant potential dynamos, that is by machines which are so constructed that they will deliver current varying in amount between zero and the limit of their capacity and at the same time maintain a pressure at the terminals that is very nearly uniform.

Though a shunt machine comes nearer to fullfilling the requisite conditions than a series, still neither are self-regulating. The shunt machine with a variable resistance in series with the field coils, has been used as described on a previous page. There are also other arrangements which could be used to attain the same end but as they have not found favor in practice they will not be considered here.

It has been shown how with the shunt machine, owing to increase in current, the volts lost in the armature are increased, resulting in decreased ability to sustain the output. To make a shunt machine self-regulating some means must be provided for increasing the field strength with the increase of the strength of the current. This can be easily accomplished by winding on the field magnets a coil consisting of a few turns of heavy wire through which the whole current of the circuit can be passed. This series coil is sometimes called the compensating coil because it compensates for the weakening tendency when the shunt coil alone is used. A machine with such a winding is called a **compound-wound machine.** Being a combination of the series and shunt machines, its characteristic curve is a combination of the two, the

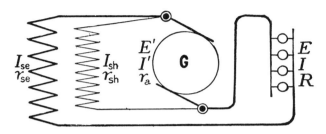

Fig. 9. Ordinary, or Short Shunt Compound Dynamo.

rise in the characteristic of the series machine occurring at the point where that of the shunt machine decreases.

There are two methods of connecting up a compound-wound machine, called respectively **short shunt** (see Fig. 9) and **long shunt** (see Fig. 10), according as the shunt terminals include the armature, that is, connected from brush to brush, or the series coil in addition, that is, connected from brush to end of series coil. It

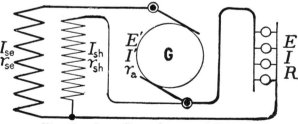

Fig. 10. Long Shunt Compound Dynamo.

makes very little difference in practical results which form of connection is employed, since the resistance of the series coils is always very small. The characteristic curve of a compound machine generally approaches a horizontal line, often it rises slightly at the beginning and then continues nearly horizontal throughout the range of loads that it is required to carry.

As the load increases and the resistance of the outside circuit decreases the main current in the series field coils increases and the current in the shunt coils still remains constant as it is now fed with a constant voltage. Thus although there are more lost volts in the armature to cut down the voltage, and although there is a much greater demagnetizing effect produced by the armature yet the voltage supplied to the external circuit is still maintained constant. A compound dynamo therefore if properly proportioned, will supply a practically constant voltage at all loads. In the case of the ordinary or short shunt compound dynamo, the potential at the brushes is kept constant. In the case of the long shunt dynamo the potential at the terminal of the working circuit is constant. The latter arrangement therefore is somewhat preferable but either arrangement proves satisfactory in well designed dynamos as the actual value of the difference in the two cases is generally very slight.

In the case of the ordinary or short shunt compound dynamo the series coils furnish the excitation required to produce a potential such as will compensate for the lost voltage in the armature and the demagnetizing effects due to the armature. In the case of the long shunt dynamo the series coils compensate for these same losses as well as for the lost voltage of the series coils.

It is advisable and generally customary to put a somewhat greater number of series turns on the field coils than is necessary to overcome armature reactions and lost voltage in order to have the dynamo give a somewhat greater voltage at full load than at no load. This process which is called **overcompounding** is calculated for a rise of voltage of about four or five per cent. As the load increases an engine often runs a few revolutions slower, or there is a trifle more slipping of the belt which causes the speed of the armature to drop a little; the increasing load is also accompanied by a greater lost voltage in the line or feeders; hence the dynamo should be overcompounded to make up for these losses. If the load is at some distance from the generator there might be considerably more than 5 % drop of voltage at full load. This drop can be figured and a machine may be especially compounded for this loss at the time it is built. The object of course is to keep the voltage constant at the lamps, and if they are some distance away and the load is constantly changing it is often necessary and advisable to run so-called *pressure wires* from the center of distribution back to a voltmeter in the dynamo room. Then if the dynamo is not overcompounded so as to give the proper pressure at all loads the pressure may be varied by adjusting the rheostat which is in series with the shunt coils of the dynamo. Even though the dynamo is properly compounded it is almost always necessary to regulate with the rheostat also in the case of incandescent lighting, especially where the lamps are distributed over a wide territory.

In designing a dynamo for any given output it is necessary to make due allowance for the total energy to be generated which is the total voltage E' multiplied by the total current I'. In order that the energy lost in the series field turns may not be too great, it is well to have them wound near the field cores so that each turn may be as short as possible and then the shunt coils may be wound outside of the series turns.

In Fig. 11 are shown some characteristic curves of a compound-wound Crocker-Wheeler dynamo running at 1500 revolutions per minute.

The upper curve is the external characteristic of the dynamo running with all the resistance cut out of the shunt field regulating rheostat. It is running, therefore, with a maximum field excitation and giving its maximum voltage, about 127.8 volts at no load. It will be noted that as the load increases the voltage drops somewhat. It is evident that the magnetic flux due to the series ampere-turns on the field coil is not great enough to make up for the armature demagnetizing effects and the lost volts in the armature. Therefore there are not enough series turns, and the dynamo is undercompounded when running at 127.8 volts.

Next some of the resistance of the regulating rheostat is put in series with the shunt field until the voltage falls to 109.2 volts. By increasing the load the voltage increases, being from about 4 % to 7 % higher as the load increases than it was at no load. Now the rise in voltage, due to compounding, makes up for the loss in the line wires and is a little too great in the case of the 7 %, and in order to keep the lamps from burning too brightly the rheostat handle would have to be turned so as to cut the voltage down a little.

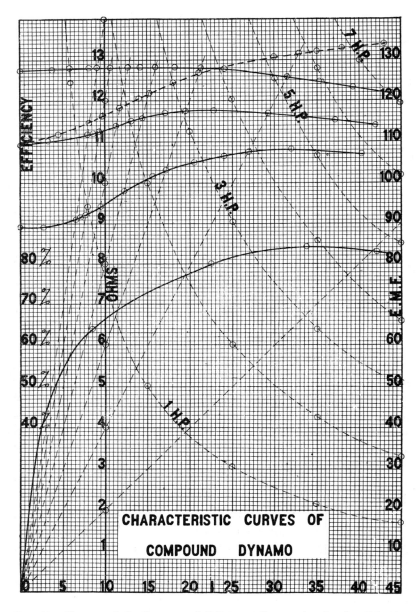

Fig. 11. Characteristic Curves and Efficiency Curve of a Crocker-Wheelei Compound Dynamo.

Judging from these two curves it appears that if the dynamo were run at a little higher initial voltage, say 112 or 115 it would be then running at just about the proper voltage for which it was compounded. The percentage of the overcompounding would then be about four or five per cent and the dynamo might run at varying loads with little or no need of adjusting the rheostat.

The curve shown by broken line (see Fig. 11) is the total characteristic curve derived by adding the values of the lost volts in armature and in series field to the value of the external voltage.

The lower compound characteristic curve starting with an initial voltage of 89 shows that the dynamo is very much overcompounded when running at this voltage. The rheostat has been so adjusted that less current flows through the field coils, the shunt current is reduced and the value of the shunt ampere-turns is reduced. On the other hand the value of the series ampere-turns is as great as formerly as the load increases, and so the value of the series excitation is now large in comparison with that of the shunt. The voltage therefore increases greatly with the load, causing the dynamo to be greatly overcompounded when started at an initial voltage of 89.

In Fig. 11, is also given a curve of commercial efficiency for this 5 H. P. Crocker-Wheeler compound dynamo, the commercial efficiency being the useful electrical work, $E\,I$, derived from the dynamo, divided by the value of the mechanical work delivered to the pulley of the dynamo.

The following table contains the data from which the characteristic curves shown in Fig. 11 were drawn.

I	E	I	E	I	E
0	127.8	0	109.2	0	89.
3.6	128.0	3.4	110.0	2.7	89.7
8.4	128.0	8.7	112.0	7.2	91.3
14.2	127.7	13.2	115.0	11.3	96.4
18.4	127.7	17.1	117.0	15.6	101.8
23.8	127.6	22.6	117.0	20.6	105.0
27.2	127.2	25.6	116.8	24.6	105.4
30.9	126.0	29.3	116.0	27.9	107.5
35.2	124.0	33.7	116.0	32.0	108.2
39.3	122.9	37.1	115.9	36.1	108.0
43.8	121.9	42.8	114.0	40.3	107.0
				40.4	107.0

The characteristic curves of a 5 K.W. Edison compound

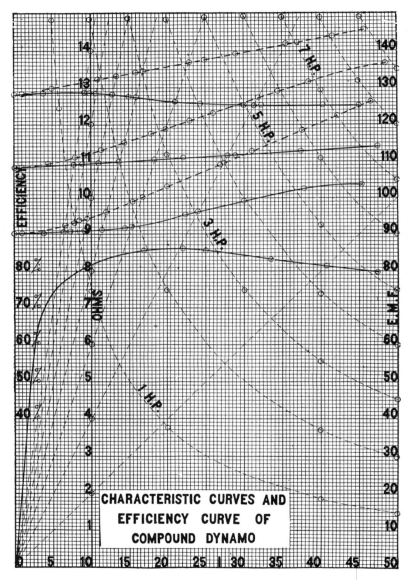

Fig. 12.   Total and External Characteristic Curves and Efficiency Curve
of 5 K. W. Compound Edison Dynamo.

dynamo are given in Fig. 12, for initial voltages of 131, 110, and 90, at a speed of 1,700 revolutions. In testing dynamos it is very difficult to keep the speed absolutely constant. There may be a variation in the speed of the main pulley from which the dynamo is run, and the belt may slip more and more as the load increases. However, as the voltage is proportional to the speed, the observed voltage may be corrected for the observed speed quite readily. The reading of the voltmeter and the tachometer should both be taken at the same instant.

The characteristics of the compound Edison dynamo as shown in Fig. 12, present very much the same phenomena as were exhibited by the characteristics of the Crocker-Wheeler dynamo in Fig. 11. The Edison dynamo, the curves of which are shown, is probably properly overcompounded for about 4 % or 5 % when running at an initial voltage of about 115. A curve of electrical efficiency is also shown.

The following table gives the data for the characteristic and efficiency curves of the 5 KW. compound Edison dynamo.

Speed.	E	E corrected to 1,700 rev.	I	Speed.	E	Corrected E	I	$\frac{E\,I}{E,I,}$
1740	131.5	128	0	1720	110	108.1	0	0
1725	130.0	128.1	3.8	1715	111	109.3	8.7	.82
1730	130.0	128.3	9.8	1690	111	111.0	22.0	.86
1725	128.5	126.7	14.7	1670	110	111.3	28.7	.87
1720	128.0	126.6	19.2	1665	112	113.1	37.3	.82
1715	126.5	125.4	24.3	1660	112	114.0	47.5	.80
1710	126.5	125.6	28.0	1730	91.5	90.0	.0	
1705	125.5	125.3	32.2	1730	91.5	90.0	7.4	
1685	124.0	125.2	36.0	1725	93.0	91.6	14.6	
1670	123.0	125.3	40.3	1715	96.5	95.7	22.4	
1665	123.0	125.6	45.0	1685	99.0	99.9	30.0	
1635	118.5	123.3	46.5	1660	99.5	101.9	38.0	
				1640	100.0	103.7	45.4	

THE MAGNETIC CIRCUIT.

Every dynamo-electric machine or electrical instrument whose working depends upon the laws of magnetic induction, must have a complete magnetic circuit just the same as it has a complete electric circuit. Moreover the laws governing the magnetic circuit are similar to the laws governing the electric circuit. The one important difference between the magnetic circuit and the electric circuit lies in the difference of materials of which the two

are composed.   Every substance is a conductor of electricity to a
certain extent; there is no perfect non-conductor or insulator.
The best conductors are metals and graphite.   Dry air, dry wood,
mica, rubber, etc., are such very poor conductors of electricity that
they are called insulators or non-conductors.

Almost all substances are fairly good conductors of magnetism.   Iron and steel in their various forms are *excellent* conductors,
and nickel, cobalt and oxygen lie between these and the fairly
good conductors in their magnetic qualities.   Fairly good conductors include solids, liquids and gases, metals and non-metals,
organic and inorganic substances; and with the exceptions named
above and a few others, they are *equally* good conductors.   Air is
generally taken as the standard and is neither a better nor a
poorer conductor than gold or copper.

The greater the current in an electric circuit the greater is
the amount of energy required to cause the current to flow, provided the resistance remains unchanged.   In like manner the
amount of energy required to produce a magnetic flow in air is
proportional to the strength of the magnetic field produced.   Iron
and steel however have induction coefficients which are very
important in their bearings.   When an electromagnet is weakly
magnetized the amount of electrical energy required is nearly proportional to the strength of magnetism.   After the metal begins
to be **saturated** with magnetic lines, the amount of electrical
energy required to increase the magnetism becomes much greater
than formerly.   After the metal has become strongly saturated
with magnetic lines even a large increase in the electrical energy
supplied may produce almost no increase in the magnetic intensity.

In order to have an electric flux or electric current in an electrical circuit it is necessary to have an electromotive force.   Similarly in order to have a magnetic flux or magnetic field it is first
necessary to have a **magneto-motive force,** (M. M. F.).   Lode-
stones and other so called permanent magnets have a magnetomotive force that keeps up a magnetic flux or field constantly.
By passing an electric current through an electric circuit a magneto-motive force is induced at right angles to the path of the
electric conductor.   In the electric circuit a given E. M. F. will
cause a larger current to flow through a low resistance conductor

than through a high resistance conductor.   Similarly with a given
M. M. F., a greater magnetic flux is produced in a good magnetic
circuit of iron or steel than in a poor magnetic circuit of air or
copper.

The geometrical distribution of the magnetic lines of force
about a conductor carrying a current is shown in Fig. 8, which
gives an end view of the conductor.   The current is assumed to
be flowing toward the observer, and the direction of the magnetic
flux is shown by arrows.
As shown by the flux
paths 1, 2, 3, 4, 5, the
magnetic density is greater
near the wire, diminish-
ing rapidly in strength
farther from the conduc-
tor.   The mathematical
law of this diminution is
as follows: the intensity
varies inversely as the
distance from the con-
ductor.

If an electric conduc-
tor is bent into the form
of a loop and an electric
current of one ampere

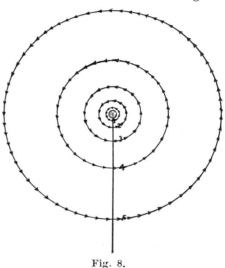

Fig. 8.

passes through the conductor then a certain number of magnetic
lines of force pass through the loop.   After following curved paths
they will return to the loop passing through it again in the same
direction as at first.   Now if two amperes flow through this conductor, or the same current (one ampere) flows through two similar
conductors laid side by side, then in either case just twice as many
magnetic lines will flow through the loop as in the first case.
Therefore the intensity of the magnetic flux through a loop composed of an electric conductor carrying a current, depends upon
and is directly proportional to the number of **ampere-turns.**
This magnetic flux can be made very great either by forcing a
large current through the conductor, or by having a very large
number of turns in the loop, or by both.

The absolute unit of current being ten times the ampere, the ampere-turn is only one tenth the value of the current-turn. Of the two, the ampere-turn is in more common use as a practical unit. The fundamental unit of M. M. F. in the United States is called the **gilbert,** and is produced by $\frac{1}{4\pi}$ of a C. G. S. unit or current-turn, or by $\frac{10}{4\pi}$ ampere-turns. Therefore by dividing the number of ampere-turns in any coil by $\frac{10}{4\pi}$, the M. M. F. of the coil is obtained in gilberts.

The quantity of magnetic flux in any magnetic circuit depends upon the M. M. F. and upon the nature and form of the circuit. In an electric circuit the current produced by an E. M. F. depends upon the resistance of the circuit. Similarly in a magnetic circuit the magnetic flux produced by a M. M. F. depends upon the **magnetic reluctance** or simply **reluctance** of the magnetic circuit. Magnetic reluctance is similar to electric resistance and is often defined as *the resistance of a circuit to magnetic flux.*

The resistance of an electric circuit measured in ohms is equal to the **specific resistance** or **resistivity** of the conductor, multiplied by the length and divided by the cross section of the wire. Similarly, the reluctance of a magnetic circuit, measured in **oersteds,** is equal to the **specific magnetic resistance** or **reluctivity** of the magnetic conductor, multiplied by its length in centimeters, and divided by its cross section in square centimeters. The reluctivity of a vacuum is unity and the reluctivity of air is generally considered equal to unity although it varies slightly. All other non-magnetic substances such as brass, copper, glass, wood, and water also have a reluctivity which is practically equal to unity. The reluctivity of magnetic metals such as iron, steel, nickel, and cobalt vary quite widely from unity and their values also vary quite widely among themselves.

The unit of reluctance by which the magnetic resistance of a circuit is measured is called an **oersted,** and is defined as the reluctance or magnetic resistance of a cubic centimeter of air measured between two opposite faces, hence **the reluctance of a** column of air 20 cm. long and 2 cm. square (4 sq. cm.) is 5 oersteds.

In the electric circuit,

Electric flux (in amperes) =

$$\frac{\text{Electromotive force (in volts)}}{\text{Resistance (in ohms)}}, \quad \text{or } I = \frac{E}{R}.$$

In the magnetic circuit,

Magnetic flux (in webers) =

$$\frac{\text{Magneto-motive force (in gilberts)}}{\text{Reluctance (in oersteds)}}, \quad \text{or } \Phi = \frac{\mathcal{F}}{\mathcal{R}}.$$

If a non-magnetic ring (see Fig.14), is uniformly wound with insulated wire, and a current is passed through the wire, the magnetic circuit will be completely closed by the coil or solenoid, and no magnetic flux will leak outside of the coil. If the cross section of the ring is 15 sq. cm., and the mean circumference is 45 cm., the reluctance of the magnetic circuit will be $\frac{45}{15} = 3$ oersteds. If the number of turns of wire is 100 and the current 5 amperes the M. M. F. in the magnetic

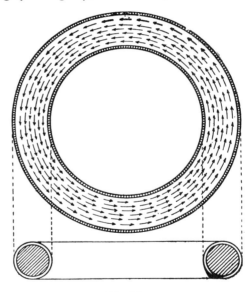

Fig.14.

circuit will be 500 ampere-turns, or $500 \div \frac{10}{4\pi} = 628.32$ gilberts. The magnetic flux in webers will be

$$\frac{628.32}{3} = 209.44 \text{ webers.}$$

The weber is the *magnetic line* of the preceding section.

The greater part of the magnetic circuits in dynamos consists almost entirely of iron.   The reluctivity of iron and steel varies greatly, according to hardness and chemical composition, and also the degree to which it has been magnetized.   When iron is fully saturated magnetically its reluctivity is as great as the reluctivity of air.   The reluctivity of soft iron or steel, when it is not strongly magnetized, is often several thousand times less than that of air. The ferric magnetic circuit is an easy one to calculate, however, since the magnetism is confined almost exclusively to the iron, thus giving a magnetic circuit of definite and easily computed length and cross-section.

There is some leakage of magnetic flux from the iron into the air.   For dynamos of ordinary sizes and shapes it varies from 15% to 30% of the flux through the field coils.   It may be separately calculated and allowed for.

The **magnetizing force,** represented by the symbol $\mathcal{K}$, may be defined as the rate at which the magnetic potential diminishes in a magnetic circuit.   The total fall of magnetic potential in a magnetic circuit is equal to the M. M. F., just as the total fall or drop of electric potential in an electric circuit is equal to the E. M. F.   The total difference of magnetic potential in the magnetic circuit shown in Fig. 14 is 628.32 gilberts.   As this circuit is symmetrical and uniform throughout, the drop of magnetic potential is uniform and equal for each unit of length.   The mean length of the circuit being 45 cms., the rate of fall of potential is about 14 gilberts per centimeter all around the circuit. This is the magnetizing force, or the magnetic force, represented by $\mathcal{K}$.   When there is no iron or other magnetic substance in the circuit, the value of the magnetizing force $\mathcal{K}$ is equal to the flux density $\mathcal{B}$.

If in any magnetic circuit the M. M. F. expressed in gilberts is divided by the length of the magnetic circuit in cms., the average value of the magnetizing force will be obtained.   Thus in a long helix, if there is a M. M. F. of 500 gilberts, and the magnetic circuit inside the helix has a length of 50 cms., the magnetizing force will be 10 gilberts per centimeter, and the flux density will be 10 gausses, provided there is no magnetic substance in the circuit.

If there is iron or any other magnetic substance in the circuit the magnetizing force will still remain 10 gilberts per centimeter although the magnetic flux density will be greatly increased according to the permeability of the magnetic circuit.

FORMS OF FIELD MAGNETS.

The function of the field of a dynamo is to provide a space filled with lines of magnetic induction which the conductors of the armature may cut, thus causing an electromotive force to be generated in them. It is immaterial to the generation of an electromotive force whether the armature or the field revolves, though generally the field is stationary. It is usually desirable to maintain the magnetic induction or number of magnetic lines at constant value. The first field magnets to be used were permanent steel magnets, and small dynamos are still built with them. The magneto-machine is a shuttle-wound dynamo having its field produced by permanent magnets. Such machines can not, of course, have as large a capacity as others equipped with electromagnets, and they are subject to loss of magnetism which after a while is liable to render them useless unless they are remagnetized. The fields of modern dynamos are supplied by electromagnets, connected to the circuit of the dynamo according to the purpose for which it is built. The different forms, named according to the winding are, — the magneto, the separately excited machine, the series machine, the shunt machine and the compound-wound machine.

The separately excited dynamo has its field coils fed from some outside circuit. This form of winding is not used much with direct current machines except for special cases. Its place is with alternate current machines where it is inconvenient to excite a machine by its own current.

Field magnets for dynamos and motors have been made in almost every conceivable shape. Electrically speaking the best form to use is one which has the most compact arrangement, presenting the shortest path for the flow of magnetism, the greatest cross section, and the fewest number of joints. Such a form would have the greatest permeability and would consequently be the most efficient. But the form which is most desirable electric-

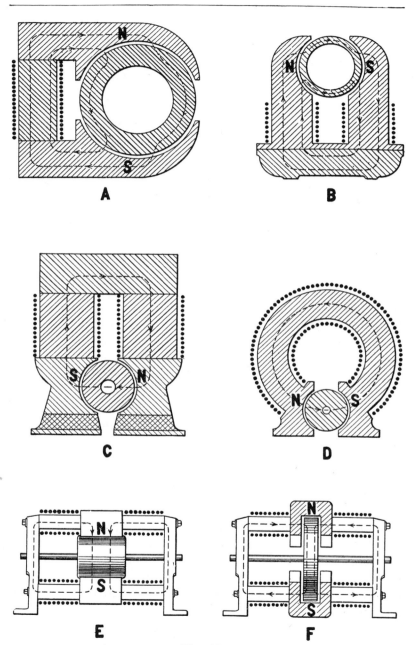

Fig. 15.

ally may often be very difficult and expensive to construct so that the forms which have become fixed in practice are usually the ones which have been found to give the best results at lowest cost.

Some of the more usual forms are shown in Figs. 15, 16 and 17. All types come under three principal heads; the single, the double, and the multiple magnetic circuit.

The horseshoe form is the type of the single circuit bipolar field magnet frame. It is placed upright, C Fig. 15, or inverted, B Fig. 15. The Edison machine is a type of the upright form, which has the disadvantage that it must be mounted on a zinc, brass or some other non-magnetic base, because if the latter were of cast iron, or any good magnetic conductor, the magnetism would be largely diverted by it instead of passing through the armature. The Thomson-Houston machines, now built by the General Electric Co., are of the inverted horseshoe type. The Jenney Electric Co. uses the horizontal type and the vertical form is employed by the Excelsior Electric Co. A modification of the horseshoe magnet type as generally used, is that employed by the Eddy Electric Co., D Fig. 15. In this machine there is no yoke and simply a half ring joins the pole pieces. This gives a very good magnetic circuit but is expensive to build. Another modification is the one magnet form, which has only a single coil wound on the portion of the frame corresponding to the yoke, as A Fig. 15, and this, like its prototype the double coil frame, is made in several positions.

The double magnetic circuit form is equivalent to two horseshoe magnets joined at the poles. The chief advantage of this type is in its greater mechanical strength. The coils are sometimes placed on the two upright sides which connect the top and bottom yokes, as shown by K. Fig 16 and P Fig. 17, the pole pieces being formed on the yokes. Again the pole pieces are elongated to make room for the coils and are placed either vertically or horizontally as shown by G and I. The double magnetic circuit type with internal poles (G Fig. 16) is called *ironclad* and is used in places where the machine might be exposed to accident. A modification of this type is used for street railway motors, the outside connecting pieces are thinned out and spread around the motor so as to completely cover it, affording perfect protection against injury from water or from mechanical means.

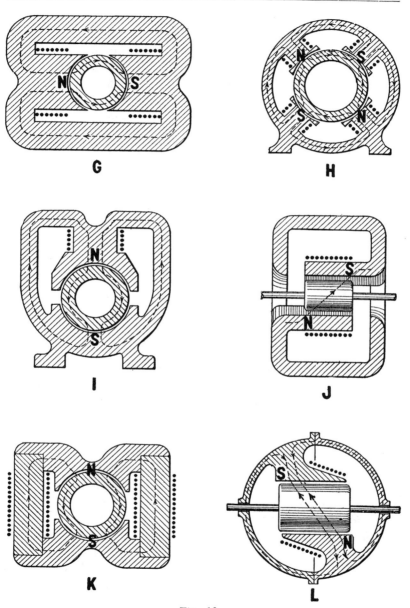

Fig. 16.

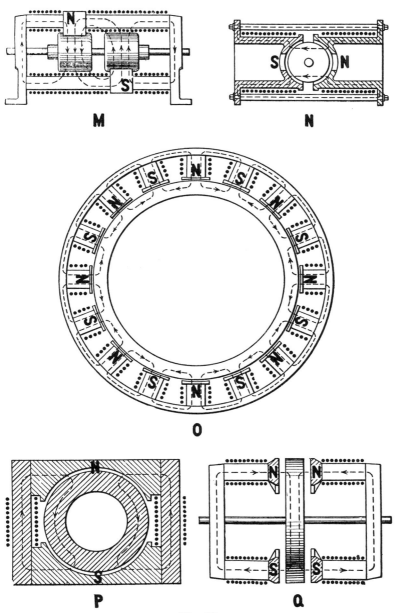

Fig. 17.

A curious modification is the Thomson-Houston "bird cage" arc light machine shown by N Fig. 17. It has tubular magnets and the pole pieces are formed to receive a spherical armature. The return circuit or yoke is made of wrought iron bars. The advantage claimed for this construction is that less wire is needed. Another ingenious form of the ironclad type is that used in the Lundell motor shown in section by L Fig. 16. There is but one coil which encloses the pole pieces, the whole being in turn completely incased by the remainder of the magnetic circuit.

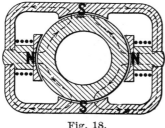

Fig. 18.

The double magnetic circuit type, modified by the formation of a pair of extra poles on the pieces which form the return circuit as in Fig. 18, gives rise to the four-pole type and hence to the multi-polar multi-circuit type.

The multipolar field magnets are usually circular in form with inwardly projecting pole pieces. Siemens & Halske use a form having outwardly projecting pole pieces with the armature ring rotating outside the field as shown in Fig. 20. This gives a better magnetic circuit but introduces mechanical difficulties in securing armature supports. Sometimes the multipolar field is built with the coils on the yokes between poles instead of on them.

The circuit of the Ball dynamo is shown by M Fig. 17, and that of the Brush dynamo by Q. Fig. 19 and H

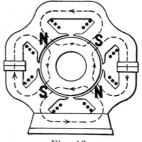

Fig. 19.

Fig. 16 show magnetic circuits of typical small multipolar dynamos. The magnetic circuit of a large multipolar dynamo with inwardly projecting poles is shown by O, Fig. 17.

For large machines the multipolar type is most in use because it results in a saving of both copper and iron and gives a more symmetrical arrangement. Below 10 K. W. capacity the bipolar forms are to be preferred as they are simpler. The bipolar single circuit types are generally superior to the double circuit types because they need less wire and are less subject to magnetic leakage.

DYNAMO CONSTRUCTION.

The Armature. The armature of a dynamo is that portion of the machine in which the electrical output is generated. The principle underlying every armature is the law of induction, that any conductor which cuts across a magnetic field has electro-motive force gener-ated in it which tends to set up a current. The various arma-tures of modern com-mercial machines are the mechanical means which experience has shown to be best adapted to carrying out the law of induc-tion serviceably and efficiently under the various conditions and uses to which the machines are sub-jected.

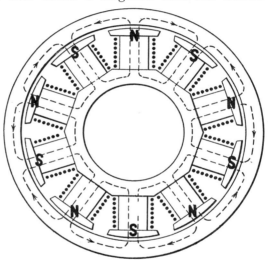

Fig. 20.

The two forms of armature in common use and from which nearly all others have been derived are the Gramme or *ring-wound* armature, and the *drum* or shuttle-wound armature. Fig. 21 shows a diagram of a simple ring armature with one coil, and Fig. 22 a diagram of a drum armature. In practice the number of coils on a single armature is of course large. In the Gramme type the iron core is a ring on which the coils are wound. In the drum type the core is a cylinder, the coils being wound on the surface and ends. Where slow speed is desired the ring armature is most generally used; the ventilation is better and a high insula-tion is more readily obtained.

There is also the pole armature, with its conductors wound upon iron cores projecting radially outwards; and the disk armature which has no core, the conductors forming a flat disk which rotates in a narrow gap between the pole pieces.

The Armature Core.   The iron armature core not only serves as a support for the armature conductors but also serves as a medium of low reluctance through which the magnetic flux from pole to pole may readily pass.

If the core consisted of one solid piece of metal, it would, since it rotates in a varying magnetic field, have currents induced in it, as is the case in the conductors on its surface.   The iron itself would form a closed circuit for these currents, which are known as Foucault or eddy currents.   Their presence would not only be a waste of energy, but would also cause injurious heating.   These currents are, however, largely reduced by laminating

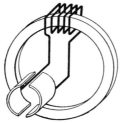

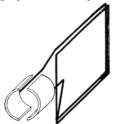

Fig. 21.   Fig. 22.

the core, that is, by building it up of thin insulated sheets of iron. These sheets or discs are perpendicular to the direction in which the induced currents would flow, and therefore break the otherwise closed circuit.   The discs, being perpendicular to the direction of the induced currents are parallel with the lines of magnetic flux and therefore this lamination does not increase the resistance to the magnetic flux.   The core of a drum armature is made up of a number of circular discs, and a ring core is made up of a number of similar rings.   The discs may be insulated from each other by sheets of paper, a coating of varnish or by rust that is allowed to form on their surface.

The higher the permeability of the material of which the core is composed, the less will be the hysteresis loss and the greater may be the magnetic flux per unit of cross section; the core discs are therefore punched out of the softest sheet iron or soft steel, which in some cases is afterwards annealed.

It has largely become the practice to use **toothed** armature cores shown in Fig 23, the conductors being laid between the teeth and within the outside surface of the core. The air-gap between the core and pole pieces can then be reduced to a minimum, and the conductors are well protected from injury. The conductors are also securely held in place. Such cores are, however, more expensive to make and unless proper precautions are taken, the teeth are likely to set up eddy currents in the pole piece surfaces.

Armature cores are usually built up on an internal frame or skeleton which is in turn firmly keyed to the shaft. In drum armatures the core discs may be keyed directly to the shaft, or the shaft and hole through the discs may be made square or octagonal. The discs are sometimes held together by large nuts screwed directly on the shaft, or, as is usually the case, by bolts passing through the discs. In the latter case end-plates somewhat thicker than the discs are provided, and to reduce the presence of eddy currents which

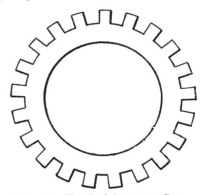

Fig. 23.　Toothed Armature Core.

would occur if the plates were of iron, they are sometimes made of brass or other non-magnetic metal. These bolts and end-plates are always insulated from the core by tubes and washers of paper, fiber, or mica.

The core of ring and of large drum armatures, being in the form of a ring, must be supported on some form of spider. This consists of a hub or sleeve keyed to the shaft and is provided with projecting arms which support the core. A simple construction is shown in Fig. 24. Two spiders are used, the core being clamped between them. It will be noticed that the bolts do not pass through the core and therefore are not in the path of the magnetic lines.

Before the conductors are applied the core should have all sharp or rough edges removed. In order to secure good insulation from the conductors the cores are covered with a few layers of

insulating material ; smooth cores have a continuous layer of insulation over their surfaces, whereas in toothed cores the insulation is
applied only in the slots between the teeth.

Armatures of bipolar dynamos usually run at a speed of from
800 to 1,400 revolutions per minute ; the armatures of direct connected machines and others which are of the multipolar type sometimes run as slow as 150 per minute, but their diameters may be so
large that the peripheral speed is
not far from that of a smaller
machine. Under these conditions
the centrifugal force acting on the
conductors is quite appreciable.

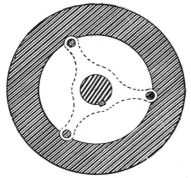

There is also another mechanical force to be provided for due to
the magnetic pull exerted on the
conductors by the field. It is
against this pull that the steam
engine driving the dynamo does
work, and were they not strongly
secured the conductors would be

Fig. 24.   Shaft and Spider Supporting Armature Core.

stripped from the armature. In order to prevent any motion of
the conductors during the operation of an armature, hard brass
wire is wrapped around the out-
side in bands ; these are called
binding wires.

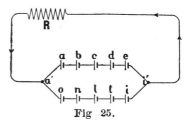

Fig 25.

On account of the high speed
and small clearance in the air-gap,
armatures are very carefully *balanced* to prevent injurious vibration. The usual method of doing
this is to support the armature and its shaft on two horizontal,
wedge-shaped metal rails called knife edges. The armature is
then given a turn and it is noted whether there is any tendency
to come to rest with any particular part of the armature down-
wards. If this is the case, small pieces of lead are inserted on
the opposite side until the armature will come to rest in any
position.

Armature Windings. The law of induction tells us that a

loop of wire carried through a field in such a manner as to vary continuously the number of lines of force that are included within its boundary will have an E. M. F generated in it. Two loops connected in series will double the effect, or twice the field strength will also double the effect. Dynamos are designed with this in view, and to produce the desired voltage the coils are made of a certain number of turns and the fields of a certain strength.

There are two general methods of winding armatures giving rise to the terms **closed coil** and **open coil.**

Closed Coil Armatures. The closed coil type of armature is used for constant potential machines, in incandescent lighting and power distribution, and in some cases for constant current machines. The coils are wound continuously around the ring or cylinder as the case may be, the end of one coil being joined to the beginning of the next, and this junction is connected with a segment of the commutator.

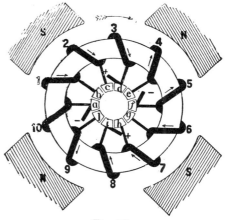

Fig. 26.

Each coil adds its quota to the electromotive force at the brushes, the amount depending at any moment on the position of the coil in the field. As the coils revolve the E. M. F. generated by some coils is increasing, while that generated by others is decreasing; but the sum total between the brushes remains the same, for the loss in effectiveness of one coil is balanced by the gain in another. The effect of the coils in a closed circuit winding is sometimes compared to a set of voltaic cells, one in each coil, as shown in Fig. 25. This would, however, be a more accurate illustration if the voltage of the cells were of varying strength according to their position in the chain. When two brushes are used there are two paths for the current from one brush to the other through the armature, and the resistance of the armature is then $\frac{1}{4}$ the resistance of the total amount of wire upon it, for the total length of wire is halved and the halves are connected in parallel.

There have been many variations in winding, all based on the drum or ring winding, and resulting from the necessity of reducing the cost of manufacture and from the increase in the number of poles used in a single machine. The armatures of small machines are usually wound with copper wire, which in larger machines is replaced by copper bars, these bars being laminated above certain sizes to prevent the occurrence of eddy currents in their own mass.

The use of more than two poles greatly increases the variation which it is possible to make in the windings. Fig. 26 illustrates the winding and connections of a four-pole ring armature. As there are

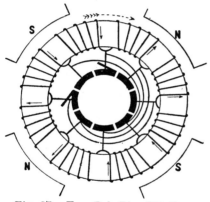

Fig. 27.   Four-Pole Ring Winding.

four neutral points it is necessary to use four brushes, two positive and two negative, the winding forming a closed circuit from brush to brush. To supply the external circuit the two positive brushes are connected giving one positive terminal, and the two negative brushes are connected giving one negative terminal. Thus the external current in passing through the machine subdivides at the two brushes and again subdivides in the armature, therefore only $\frac{1}{4}$ of the total current passes through each conductor. The internal resistance is $\frac{1}{16}$ of that of all the wire on the armature. Similarly more poles would cause the current to divide a greater number of times.

It is not always necessary to have as many brushes as there are neutral points or poles, although this is usually the case. By the simple device of **cross-connecting,** a diagram of which is shown in Fig. 27, the number of brushes may be reduced to two. Each coil is cross-connected with others occupying a similar position under the corresponding pole. This cross-connection may be either in the armature itself or in the commutator.

A method by which armature windings and connections are clearly shown consists in laying the poles and winding out in a plane surface, or in showing the winding as it will appear when

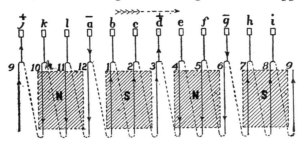

Fig. 28. Developed Ring Winding.

unfolded upon a plane surface. Such a development of the winding of a ring armature is shown in Fig. 28. Figs. 29 and 30 show two windings called the **lap** and the **wave winding** respectively. They are more like the drum winding than the ring, being entirely exterior to the core. A portion called an *element* is shown heavier than the remainder of the windings which are all similar to it.

A conductor on a revolving armature generates an E. M. F. under a north pole in the opposite direction to one under a south

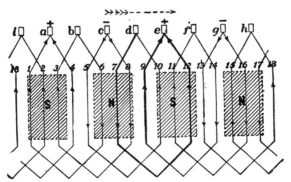

Fig. 29. Developed Lap Winding.

pole, so that it is possible to lay out a wire on the surface of an armature, carrying it across the ends from point to point, in such a way that it will be subjected to an inductive effect tending to drive the current in the same direction throughout its length. The two ways of obtaining this result are illustrated in Figs. 29 and 30. The reason for calling them *lap* and *wave* windings is that one doubles back on itself every turn, while the other goes ahead in a continuous zig-zag, as will become apparent by following an element.

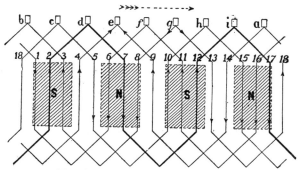

Fig. 30.　Developed Wave Winding.

Open Coil Armatures. In the open coil type of armature, which is used chiefly for arc lighting or constant current machines, the various coils, which are few in number, are not interconnected; each winding performs its function independently of the others, and is cut into the circuit by the commutator at the time of greatest activity and is entirely disconnected from the circuit when it is idle.

The commutators of such machines are made up of few segments which are insulated from each other by air spaces. These machines generally spark considerably at the brushes on account of the actual break in the circuit of the separate armature coils.

Fig. 31 shows a diagram of a Brush arc machine which is ring-wound. There are four coils, the two coils A and C of one winding being in series and terminating in a two part commutator, while the others B and D, also in series, terminate in another two part commutator. The commutators are placed side by side and the segments are made long enough to overlap so that sometimes the brushes are connected to one pair of terminals, sometimes to the other, and sometimes to both at once. When the brushes E E touch both sets of terminals at once thus bringing the two windings in parallel, a current would be set up in one winding by the other owing to the difference in activity of the two, were it not for the high self-induction of the coils. The contact between the brushes and a pair of terminals breaks just when the coils con nected have reached the point of least activity. The completed

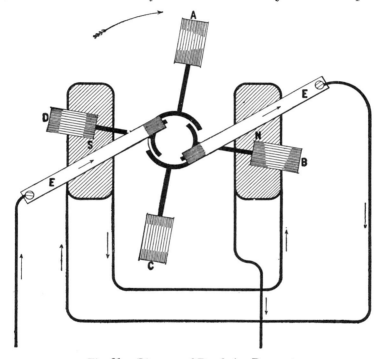

Fig. 31. Diagram of Brush Arc Dynamo.

dynamo has another set of coils between those shown, on the same ring, and another commutator and pair of brushes beside the first set. One brush of the first set is connected to the outside circuit, the other is connected to one of the brushes of the second set while the remaining brush is connected to the other terminal of the outside circuit through the field coils. In this way the two sets of windings are brought in series with each other and owing to their relative positions, one is at a maximum activity when the other is at a minimum, one increases as the other decreases, and *vice versa*, so that the sum total of the action tends to keep the current from fluctuating.

Commutators. It has been shown how current generated in a coil revolving in a magnetic field is necessarily alternating, being generated in one direction through the coil when the coil is rising through the field and in the opposite direction when it is descending. The first machines built were alternating current machines, but as the alternating current could then only be used for a few purposes the invention of the commutator soon followed. The simplest form of dynamo is a single coil revolving in a field, the terminals of the coil ending in two pieces of metal which form the halves of a circle, and with which the brushes are in contact. As the coil rotates, first one terminal and then the other comes

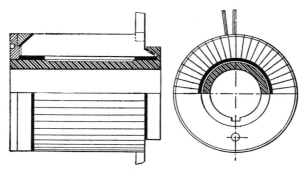

Fig. 32. Commutator.

under a brush. If the brushes are put in the right place the terminals will always have the same polarity when they make contact with the brushes so that a continuous current can be drawn from them. The current is continuous however, only as regards direction for the strength of the current is varying.

As the coil moves into the denser part of the field the current increases, and as the coil moves out the current decreases. If we had two coils at right angles to each other, moving through the field in the same time, the pulsation would only be half as great as with one. We can therefore make the pulsation as small as we like by increasing the number of coils and hence the number of commutator segments. The fluctuations become perceptible in a two pole machine when less than 30 coils are used; and the commutator bars seldom exceed 100 for this type of machine on account of increased expense. In larger machines the total number of commutator segments increases with the number of pairs of poles.

The commutator must be well designed mechanically as well as electrically. Each segment must be carefully insulated from adjacent ones and from the shaft upon which it is built up. The insulation between the segments consists of sheets of mica which have been selected for firmness, freedom from cracks, and uniformity in thickness. The segments taper toward the center so that they will build up into a cylinder. The whole is firmly clamped together by insulating rings held in place by metal rings and lock nuts. Figs. 32 and 33 show two different methods of securing the

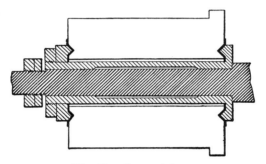

Fig. 33.   Commutator.

assembled commutator bars. The best material for the commutator bars is hard copper or silicon bronze which should be as nearly pure as possible and uniform in its structure. The commutator is designed larger than necessary so as to allow for the continual wearing away of its surface ; and also to allow it to be turned down from time to time when the surface becomes rough as the result of uneven wear or accident.

Brushes. The cross section of the brush must always be large enough to carry the maximum current which it is likely to be called upon to furnish, with a liberal factor of safety. The thickness at the contact with the commutator should not be less than one and a half times the insulation between the bars and is usually greater. The material must be softer than that of the commutator so as to wear the brush rather than the commutator.

There is a large number of different forms of brushes, each having its special advantages and disadvantages. In the Brush arc dynamo the brush consists of a simple strip of sheet copper having slits in it to insure good contact, and is set almost tangent to the commutator.

The wire brush has been much used and is still employed; it consists of a bundle of fine copper wires all soldered together at the end furthest from the commutator, and being inclined to its surface, the brush has the contact end beveled to fit the commutator in order that the ends of all the wires may make contact.

Copper gauze brushes consist of sheets of fine copper gauze folded or rolled up and pressed into rectangular form. These brushes owing to their soft spongy nature secure excellent contact with the commutator.

The carbon brush is now used more than any other. It is usually applied radially giving the advantage that if the machine is run backwards, through accident or intention, there is no danger of ruining the brush by causing it to trip on the edge of some commutator bar that might be slightly out of place. It also wears a better surface on the commutator and the amount of wear is much reduced. The collection of carbon dust is far less objectionable about a commutator than that of copper dust being less likely to cause a short circuit. The main advantage of the carbon brush is the reduction of sparking. As carbon has a high specific resistance it tends to keep the current down when the commutator bars are short circuited by coming under the brush, and there is therefore little or no sparking when a bar breaks contact with the brush. This advantage becomes less marked in machines having many segments in the commutator and very low difference of potential between them, for in such machines there would be little sparking even with copper brushes having high conductivity.

Brush-Holders. The brush must be held rigidly in position by a mechanism called the **brush-holder** which is capable of being revolved part way around the shaft so as to shift the point of contact between the brush and the commutator. The holder is usually mounted on a collar which in turn is mounted on a hub formed by a projection from the journal box. An insulated handle is provided by which to shift the brushes to the desired position, and it is often made to screw into the collar in such a way that the brush-holder can be locked in any position. In very

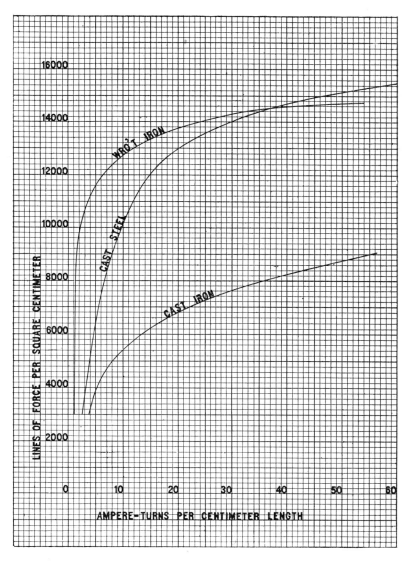

Fig. 34.   Curves of Magnetic Induction.

large multipolar machines where there are as many brushes as poles, a substantial ring is formed separate from the machine.   This ring is supported from the field casting by arms, and is shifted by the aid of a hand wheel and worm meshing with teeth on the ring.

Except in very small machines brushes are subdivided into two or more parts pivoted on the same shaft of the brush-holder. By this means of construction one brush at a time may be lifted and examined without breaking the circuit or interfering with the operation of the machine.   The brush-holders are generally provided with adjustable springs which give the brushes a proper pressure on the commutator.   If much of the current is allowed to pass through the springs, they are heated and soon lose their temper, and they cannot then exert sufficient pressure on the brush. The springs are therefore insulated, or arranged so that only a very small part of the current passes through them.

Field Magnets.   In the construction of the field magnets the principal materials used are cast iron, cast steel, and wrought iron. Usually a combination of two of these is used.   In Fig. 34 are shown curves of magnetic induction for cast iron, wrought iron and cast steel.

Cast iron has the lowest permeability of the three and hence for a certain magnetic flux must necessarily have a larger cross section.   It is customary to allow a density of about 45,000 lines per square inch for cast iron.   This material is used, on account of its cheapness, for yokes, polepieces, bases, or in general where extra weight or size is not objectionable.   This extra weight is often an advantage since it gives greater firmness and stability.

Wrought iron is generally considered to have the highest permeability of any material and a density of about 90,000 lines per square inch is usually allowed.   Owing to its high permeability. wrought iron is used where a small cross section is desirable, and this is especially the case in field cores.   To wind a certain number of turns upon a wrought-iron core will require a considerable less length of wire than if a core of some other material large enough to carry the same magnetic flux were used.   If a cast-iron core were used the length of wire would be about 1.5 times that needed for a wrought-iron core for a given number of turns.

The improved and cheaper processes of casting soft steel having a low percentage of carbon, has led to its adoption for field magnets in many dynamos. Cast steel has a very high permeability and some varieties have a higher permeability than that of wrought iron; it is customary however to allow a density of about 80,000 or 85,000 lines per square inch. Cast steel is always used for field magnets where lightness and small size are desirable.

Field Coils. For convenience in construction and repair of field coils they are usually wound on a separate form and then placed in position on the cores. The cylindrical form upon which they are wound often consists of tubes of brass or tin, having flanged ends. The coil should be insulated from this tube by a few layers of thick paper or fibre. The finished winding should be perfectly symmetrical or as nearly so as possible. The leads from the coil should be carefully protected and insulated; this is especially the case with the inner terminal, which is usually carried out through a hole in one end of the tube or spool.

As the conductors of series coils often must be of large size, they are difficult to wind, and therefore usually consist of several smaller wires connected in parallel, or of ribbons or strips of sheet copper. In compound-wound machines the series and shunt windings may be separate or on the same spool. In the latter case the shunt coil and series coils are wound in separate compartments of the spool. When the coils are separate there is the advantage of greater ease in repair and also greater radiating surface.

SPARKING.

It sometimes happens that heavy sparking takes place at the brushes of a dynamo and if this is not prevented it soon burns and roughens the surface of the commutator and the brush so badly that the dynamo is thrown out of service. Sparking may result from either mechanical or electrical faults. If one bar of the commutator is higher or lower than the rest there will be a spark every time it passes under the brush, or if the lead from an armature coil to one of the commutator bars should become disconnected, or if there should be a break in some other part of the

coil, there will be flashing when the bar passes under a brush.   If
the sparking is continuous and not due to mechanical causes
the brushes should be rocked one way or the other until a position
is found where the machine runs sparklessly, or nearly so.

When a dynamo is in operation the current passes from
the positive brushes through the external circuit to the negative
brushes, and then sub-divides and passes through the armature
to the positive brushes.   To understand the production of sparking due to an incorrect position of the brushes it is necessary
to consider the commutation at one brush only.

Referring to Fig. 35, it is evident that the current passes
through coils 2 and 3 to segment d in one direction and through
coils 5 and 4 to segment d in the opposite direction.   From segment d the current passes to the positive brush and to the external
circuit.   Further movement of the armature in the direction of
rotation will bring both segments c and d under the positive
brush at the same time and coil 3 will then be short circuited.

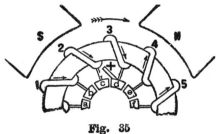

Fig. 35

That is, these two segments,
the brush and coil 3 will form
a complete circuit of very low
resistance, and the current
may pass from coil 2 directly
to segment c and the brush,
and from coil 4 directly to
segment d and the brush.
As the armature continues
to rotate, coil 3 will come under the north pole and segment c
under the positive brush.   The current must then pass from coil 4
through coil 3 to segment c and the brush.   In passing the brush
coil 3 is therefore at first short circuited and afterwards is trav-
ersed by a current in the reverse direction to that originally considered.   Each coil successively passes through the same positions.
Now, suppose the position of the positive brush is such that when
coil 3 is short circuited it is cutting no lines of force and therefore
generating no E. M. F.   Then there will be no current and as the
coil passes to the right it will be perfectly idle, and when segment
d leaves the brush the large current from coil 4 should pass
through coil 3 to get to segment c and the brush.   On account of

the large self-induction of the coils, however, the current cannot immediately rise to its full value in coil 3 and hence will prefer to jump from segment d to c, causing sparking. Now consider the result when the brush is shifted to the left or backwards, so that when the coil is short circuited it is still cutting lines of force. The generation of a slight E. M. F. will cause a large current to flow on account of the very low resistance of the coil, and the direction of this current will be opposite to that which the coil must next convey. For such a position of the brush excessive sparking will result.

Now consider the brush to be moved forward or to the right, so that while the coil is short circuited it is passing into the field at the right or under the north pole. The E. M. F. generated by coil 3 will then cause a current to flow which will be in the same direction as that in coil 4, and when segment d passes from under the brush the current in 3 already has a certain value. The correct forward position of the brush is such that coil 3 generates just sufficient E. M. F. to produce a current equal to that which passes in coil 4. No sparking will occur under these conditions, as the effect of self-induction has been counteracted and the current from coil 4 will readily pass to 3. If the brush is too far forward, too large a current will be generated, and cause sparking. The brush must be set forward a greater amount for large armature currents, since the short-circuited coil must generate a higher E. M. F. to produce the large current. The distortion of the field by the armature also necessitates the shifting of the brushes forward.

INSTALLATION AND OPERATION.

A good foundation is the first requisite in a dynamo installation. With belted machines the high speed may give trouble if the foundation is insecure. With direct driven machines the engine and dynamo should be mounted on the same bed plate if the size permits. In selecting a position for the dynamo, freedom from dampness and dust should be secured and also good ventilation. A belted dynamo should be far enough away from its driving pulley to allow the belt to have good contact. A modern direct connected unit of 15 K. W. capacity is shown in Fig. 36.

Most machines of modern design are furnished with self-aligning and self-oiling bearings. If there are a number of machines in one installation each of which is fitted with oil cups, it may be advisable to have a system of oil pipes arranged to dis-

Fig. 36.   Fort Wayne Direct-Connected Generating Set.

charge oil through a stop cock into the oil cups. All oil cans should be of non-magnetic material, to prevent accidents which might arise from the attraction that would be exercised on an iron can when placed near some parts of the dynamo.

All electrical machinery should he kept scrupulously clean.

Brushes should at all times be kept in good contact with the commutator, with their faces properly bevelled, and if they are of metal, the armature should never be allowed to run backwards. The commutator must be kept smooth and polished, as any roughness will cause sparking and a burning of the surface. If each brush-holder supports two or more brushes as is generally the case, they should be staggered enough to prevent uneven wearing of the commutator. The commutator is the most delicate part of a dynamo and consequently requires special care. Commutators should be kept free from oil, as it will carbonize and cause flashing and may even partially short circuit the bars. Where carbon brushes are used the carbon dust should be wiped off from time to time. Some lubricant is generally necessary for the commutator, and it may be applied by holding against it a coarse cloth folded smoothly and containing the lubricant. Vaseline is generally used.

The position of the brushes should be such that the machine runs sparklessly and when a change of load causes sparking the brushes should be shifted at once to the right position, as the commutator will become so badly roughened if allowed to run long in this condition, that it will have to be turned down with a cutting tool. The term *sparkless* means that there are no sparks present sufficiently large to burn the metal. When a machine is running it is possible by looking tangentially to the commutator at the brush to see a line of these blue sparks under the brush.

Dynamos as a rule are not liable to many mishaps if they are properly built and if they receive proper care, but it is often difficult to locate certain faults. A new machine will sometimes refuse to generate when first started; this is often due to entire absence of residual magnetism, making it impossible for the machine to build up. This may be remedied by exciting the field from an outside source, either another machine or a battery. It is however more likely to be due to some bad connection, such as would be produced by getting varnish under a terminal washer or by a loose wire. Sometimes the terminals of the field coils are reversed so that they oppose each other, or perhaps the brushes do not make sufficiently good contact. A shunt machine will not build up if its terminals are short circuited.

Burn-outs in the armature are sometimes produced by exces-

sive heating resulting from overload. A short circuit will very
often throw the belt or stop the engine so quickly that there is
not time for the armature conductors to melt. Continued over-
loads not great enough to throw the belt may char the insulation
until two conductors lying side by side and differing from each
other in potential, come into contact. Vibration due to a badly
balanced armature or to an insecure foundation may produce the
same trouble by causing abrasion of the insulation on the conductors, so that they touch each other or are both touching the iron
core. Dirt between two adjacent commutator bars, due to the
collection of copper and carbon dust, may short circuit the coil
having terminals in these bars sufficiently to cause it to burn out.
Sometimes the connecting pieces between the armature coils and
the commutator segments become broken which unbalances the

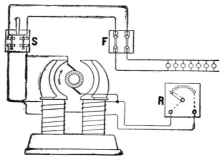

Fig. 37. Diagram of Connections of a Compound-Wound Dynamo.

armature electrically, and causes sparking at the dead commutator
bar.

The connection between a dynamo and its outside circuit
should always be made through a double-pole switch which cuts
both terminals from the circuit. A diagram of the necessary connections and wiring of a compound-wound dynamo is shown in
Fig. 37. The leads from the external circuit are first connected
to the fuses F in order to protect the dynamo from large or
dangerous currents. If a current greater than a safe one for the
dynamo passes through these fuses they melt and so break the
circuit. From the fuses leads connect with the main switch S,
and from this to the brushes through the series coils. The
rheostat R is connected in series with the shunt coils for the pur-

pose of regulating the field strength and hence the voltage of the machine. By moving the arm of the rheostat, more or less resistance is inserted or cut out from the circuit and therefore the current in the field coils is varied.

In most stations supplying electrical energy it is not desirable to have a single machine capable of taking the whole load of the station, for at low loads the efficiency would be small. It is therefore desirable to have several smaller machines, so that as the load of the station increases they can be successively added, and so always operated at full load when they are most efficient. The several machines are then connected in parallel between the bus bars from which the external circuits lead.

With shunt machines no trouble is experienced when several are running in parallel. Care must be taken before connecting a new machine to the circuit that it is built up to the same voltage as that carried on the bus bars, for if it is lower the higher voltage will send a current to the machine of lower voltage and drive it as a motor.

With series or compound machines the case is somewhat different. A conductor called an equalizer must connect together all the brushes from which the series coils start. Otherwise any tendency in one machine to raise its potential above that of the others would cause increased potential at the brushes and hence an increased current through the shunt coils, which would still further raise the potential; and if the potential became high enough it would reverse all the other machines on the circuit. This action is entirely obviated by the use of the equalizer, for if the potential of any machine should tend to rise it would drive some of its current across the equalizer connection and through the series coils of the other machines thus raising their potential with its own. Machines differing widely in size can be run together in this way if all are wound for the same potential.

TESTING.

When machinery has been installed it is customary to subject it to certain tests to determine whether or not it fulfills all the requirements. No test can be applied to determine the life of the machine. The life depends upon the general design, material of which it is made,

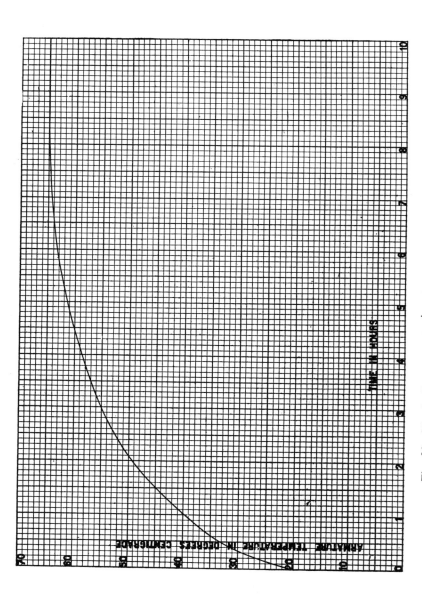

Fig. 38. Rise in Temperature of an Armature of a Large Machine.

care used in its construction, and the care it receives during its operation. Only previous experience with similar machinery is of any value as a guide in this respect.

With proper care and good materials the only defect that is likely to shorten the life of the dynamo is overheating. It is customary to specify that a machine shall be so designed that it may run continuously at full load with a rise of temperature in the armature conductors not exceeding 40° C. above the surrounding air. The commutator may be allowed to reach a slightly higher temperature, say 5° or 10° greater. The effect of overheating is to gradually char and weaken the insulation. This test is easily made after the dynamo has been in operation for a few hours at full load by placing thermometers in the armature windings at the back and front where the windings are accessible and allowing them to remain there about ten minutes covered with waste. In general the above requirement is well within the limits of safety. Fig. 38 is a curve of the rise in temperature of an armature when in operation.

In electric light supply it is very essential that the voltage at the lamps should remain at its proper value. If it is too high the lamps will be unpleasantly bright and their life will be much shortened ; if too low the lamps will burn too dimly. If short, quick variations occur, even though very small the effect will be noticed in the lamps as flickering. Hence, it is desirable that the engine driving the dynamo should regulate its speed with extreme closeness. Flickering is avoided by the use of a sufficiently heavy fly-wheel, whose proportion will depend on the speed, point of cut off, and length of stroke. It is usual to specify that the variation in speed of the combination of engine and dynamo from no load to full load shall not exceed 3 %, and the change in speed during each stroke of the engine must not exceed one per cent so that flickering in the lamps will not be noticeable.

A characteristic curve should always be taken as well as a speed curve, and if the machine is over-compounded to keep the voltage constant at some particular point on the feeder system, a curve for this point should be constructed by subtracting the volts lost in the mains from the external characteristic curve.

Multipolar machines can be built to run sparklessly under sudden variations in load as great as from no load to full load, and tests should be made by repeatedly throwing on and off certain proportions of the load, and noting the amount of sparking. It is need less to say that under a steady load the machine should run sparklessly. In many machines the position of the brushes depends somewhat on the load and requires shifting with the load.

If an insulation resistance test is made, it should be done when the machine is hot, and a source of electrical energy at considerably higher voltage than that for which the machine is built should be used, generally from 5 to 10 times as great. One pole is applied to the iron core and the other to the armature winding; if the insulation resists this test, no puncture being formed, it is without doubt free from any defect.

The efficiency of any machine is expressed by the ratio of the energy that it gives out, or output, to that which is supplied to the machine or input. This ratio multiplied by 100 gives the percentage efficiency. If the efficiency of a dynamo is 85 %, then 85 % of the mechanical energy supplied to the machine at the pulley, will reappear as available electrical energy at the terminals, while 15 % is lost. The losses are both mechanical and electrical. The mechanical losses are simply the result of friction in the bearings and brushes besides air friction due to the revolving armature. The electrical losses are due to the resistance of the armature and fields, being equal to the square of the current multiplied by the resistance. There are also additional electrical losses from eddy currents and hysteresis.

To determine the total or commercial efficiency there are several methods, mechanical and electrical, in use. If the machine is run from a belt and shaft a dynamometer can be used. There are many different forms of these machines in existence, the principle underlying them being to pass the pull of the driving power through some form of spring which is bent proportionately, the amount of the force exerted being indicated by a pointer on a scale which can be graduated to read in pounds pull. Knowing the pull applied at the circumference of the dynamo pulley, its radius, and the number of revolutions per minute, the horse-power supplied to the dynamo is found by the formula

$$\text{H. P.} = \frac{2 \pi n F r}{33,000},$$

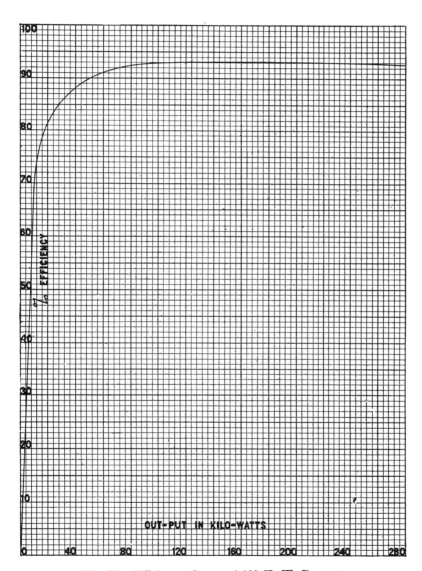

Fig. 39. Efficiency Curve of 200 K. W. Dynamo.

where n is the number of revolutions per minute, F the pull in pounds at the circumference, and r the radius of the pulley in feet.

Knowing the horse-power supplied and the horse-power taken out as given by the product of the current times the terminal voltage divided by 746, the commercial efficiency of the machine in per cent is obtained by multiplying the ratio of the latter result to the former by 100.

The efficiency is of course very much lower at light loads than at full load because friction is practically a constant quantity for a machine running at constant speed, so that the energy consumed by it bears a much greater proportion to the input at low loads than at high.

An efficiency curve of a Crocker-Wheeler direct connected multipolar 200 K. W. machine for lighting purposes is shown in Fig. 39.

It is not always possible to use a dynamometer, as for instance in the case of a direct connected machine. Under these circumstances the best method is to drive the dynamo as a motor with no load, and note the current required to just keep the machine moving at normal speed. This current multiplied by the voltage at the terminals gives the wasted power. The losses at full load are greater since the current is greater, and the $I^2 R$ losses then have a larger value.

With direct current machinery it is sometimes convenient to take indicator cards on the steam engine from which the steam power developed can be calculated, and at the same time to note the electrical output at the switchboard, the ratio of the two giving the efficiency of the combination. This procedure is only permissible when the load is quite steady; any attempt to do this on a railway generator in service is out of the question, because owing to the fly wheel sudden changes of load are not felt simultaneously in the engine and dynamo. A momentary drop of 30 % or 40 % in the electrical load would probably not reduce the indicated horse-power in the engine at all for a short space of time; the power thus liberated would go into speeding up the fly wheel.

The best method for obtaining a steady load on a dynamo is to pass the current through a water resistance or rheostat consisting of a wooden tank filled with water in which hang two iron plates. The terminals of the dynamo are connected with the iron plates, and the resistance and consequently the load is varied as desired by bringing the plates closer together or moving them farther apart. An arrangement which answers the purpose very well can be made from an ordinary barrel with 4 positive plates and 4 negative plates 12″ x 24″ and $\frac{1}{16}$″ thick, held apart by narrow boards 1″ thick placed between them. These boards should be made longer than the plates, so that they can be nailed together above and below. Ordinary furnace grates will also do very well.

Regulation can be secured by suspending the plates by some means above the barrel. They may then be lowered into the barrel to any desired depth. A cold-water supply should be furnished at the bottom of the barrel and an overflow pipe at the top. In this way the water can be carried away as fast as it is heated by the passage of the current. This apparatus will take care of 100 amperes at 550 volts. By adding a handful of salt to the water its capacity can be increased to 300 and even 400 amperes, but at such high values the ebullition is likely to become so great as to be unmanageable.

DYNAMO FORMULAS.

The complete analysis of the dynamo would be out of place here, but several important formulas will be developed to indicate the method.

For the *shunt dynamo* the analysis is as follows:

Let I' = total current in armature;

I_{sh} = current flowing in shunt coils;

I = useful current flowing in outside circuit;

E' = total E. M. F. generated in armature;

E = potential of dynamo, both for shunt coil and outside circuit;

r_a = resistance of armature;

r_{sh} = resistance of shunt field coils;

R = resistance of outside or useful circuit;

π_e = electrical efficiency.

These various quantities can be expressed in terms of each other in a number of different ways. This is of value in studying and understanding the characteristics of the shunt dynamo, and in finding the values of unknown terms when others are known. It is obvious by looking at diagram of Fig. 5 that the total current is the sum of the outside current and shunt current, or

$$I' = I + I_{sh} \text{ or } I' = I + \frac{E}{r_{sh}}$$

This latter equation shows what amount of the armature current is lost in the shunt field. Also the total current in the armature is equal to the total E. M. F. divided by the total resistance,

$$I' = \frac{E'}{r_a + \dfrac{R \times r_{sh}}{R + r_{sh}}}$$

because the joint resistance of the shunt and outside circuits (see section on the *Electric Current*) is equal to the product of the two divided by the sum of the two.

The relation can be expressed in still another way, since

$$\frac{E}{R} = I \text{ and } \frac{I}{I_{sh}} = \frac{r_{sh}}{R} \text{ from which } I' = \frac{E}{R} \left\{ 1 + \frac{R}{r_{sh}} \right\}$$

E, R, and r_{sh} are often known quantities from which I' and E' may be found by the aid of these relations. The value of the useful current in the outside circuit is

$$I = \frac{E}{R}$$

and the value of the current in the shunt circuit is

$$I_{sh} = \frac{E}{r_{sh}}.$$

The value of the total E. M. F. generated by the dynamo is

$$E' = E + I' r_a$$

from which it may be seen by inspection that the lost volts in the armature $E' - E$, increase directly with I'. If E' and r_a are considered as remaining constant, then an increase in I' increases $I'r_a$ also, and E must become less. If E is decreased and E' is constant then their difference which is equal to the lost volts in the armature must increase.

As $I' = I + I_{sh}$, we have by multiplying by r_a and substituting for $I'r_a$ its value $E' - E$,

$$E' = E + Ir_a + I_{sh}\, r_a = E + \frac{E}{R}\, r_a + \frac{E}{r_{sh}}\, r_a$$

$$= E\left\{ 1 + \frac{r_a}{R} + \frac{r_a}{r_{sh}} \right\} = E \times r_a \left\{ \frac{1}{r_a} + \frac{1}{R} + \frac{1}{r_{sh}} \right\}$$

The **electrical efficiency** of the dynamo is equal to the value of the useful energy divided by the total energy developed in the armature, the total energy itself being equal to the useful energy plus the energy lost in the armature, plus the energy lost in the field coils. Therefore the electrical efficiency is

$$\eta_e = \frac{E\,I}{E'\,I'} = \frac{I^2\,R}{I^2\,R + I'^2\,r_a + I^2_{sh}\,r_{sh}}$$

As $I' = I \times \dfrac{R + r_{sh}}{r_{sh}}$, and $I_{sh} = \dfrac{I\,R}{r_{sh}}$ we obtain

$$\eta_e = \frac{I^2\,R}{I^2\,R + \left\{ I \times \dfrac{R + r_{sh}}{r_{sh}} \right\}^2 r_a + \left\{ \dfrac{I\,R}{r_{sh}} \right\}^2 r_{sh}}$$

$$\eta_e = \frac{I^2\,R}{I^2\,R + \left\{ \dfrac{I^2\,R^2 + 2\,I^2\,R\,r_{sh} + I^2\,r^2_{sh}}{r^2_{sh}} \right\} r_a + \dfrac{I^2\,R^2}{r_{sh}}}$$

$$= \frac{1}{1 + 2\,\dfrac{r_a}{r_{sh}} + \dfrac{r_a}{R} + \dfrac{(r_a + r_{sh})\,R}{r^2_{sh}}}$$

By a mathematical analysis it can be proven that the value of η_e is a maximum when

$$R = r_{sh} \sqrt{\frac{r_a}{r_a + r_{sh}}}$$

This equation shows what the resistance of the outside circuit of a dynamo should be, knowing the shunt and armature resistances, in order to have the machine carry the proper load for maximum electrical efficiency.

Replacing R by its value in terms of r_a and r_{sh} we have the terms

$$\frac{R\,(r_{sh} + r_a)}{r^2{}_{sh}} = \frac{r_{sh} + r_a}{r_{sh}} \sqrt{\frac{r_a}{r_a + r_{sh}}} = \frac{\sqrt{r_a\,(r_{sh} + r_a)}}{r_{sh}}$$

and $\dfrac{r_a}{R} = \dfrac{r_a}{r_{sh}} \sqrt{\dfrac{r_a + r_{sh}}{r_a}} = \dfrac{\sqrt{r_a\,(r_a + r_{sh})}}{r_{sh}}$

The equation of the maximum efficiency of a shunt dynamo becomes

$$\eta_{e.Max.} = \frac{1}{1 + 2\,\dfrac{\sqrt{r_a\,(r_a + r_{sh})}}{r_{sh}} + 2\,\dfrac{r_a}{r_{sh}}}$$

The value of the armature resistance compared with the value of the shunt resistance is always so small that $r_a + r_{sh}$ may be written r_{sh} without any appreciable error and the value of $\dfrac{r_a}{r_{sh}}$ is so small that it may be neglected, then the approximate value of the electrical efficiency of the shunt dynamo may be written in very simple terms and may be considered as equal to

$$\eta_e = \frac{1}{1 + 2\,\sqrt{\dfrac{r_a}{r_{sh}}}}$$

and the approximate ratio of the shunt resistance to the armature resistance is expressed by the formula

$$\frac{r_{sh}}{r_a} = \left\{ \frac{2\,\eta_e}{1 - \eta_e} \right\}^2$$

The following table from Wiener gives the necessary ratios of the resistance of the shunt field to the resistance of the armature for various electrical efficiencies for shunt dynamos. As these values are not given in ohms but only as ratios they hold good for all shunt dynamos, large or small. It will be noted that this electrical efficiency may be as near 100% as is desired, from a theoretical standpoint, but the limit is generally governed by the practical consideration of the cost. If the armature resistance is made too small, more copper is used to carry the current than is required, and if the field coil resistance is made extremely high the wire will be very small and very expensive.

TABLE GIVING RATIO OF SHUNT TO ARMATURE RESISTANCE FOR VARIOUS ELECTRICAL EFFICIENCIES FOR SHUNT DYNAMOS.

Electrical Efficiency.	Ratio of Shunt to Armature Resistance.	Electrical Efficiency.	Ratio of Shunt to Armature Resistance.
$100 \times \eta_e$	$\dfrac{r_{sh}}{r_a}$	$100 \times \eta_e$	$\dfrac{r_{sh}}{r_a}$
80%	64	96	2,304
85	128	97	4,182
90	324	98	9,604
93	706	99	39,204
95	1,444	99.5	158,404

The *commercial efficiency* is the ratio of the output to the energy supplied; or

$$\text{commercial efficiency} = \frac{\text{useful power}}{\text{total power}}$$

$$= \frac{E\,I}{E\,I + I_{sh}^2\;r_{sh} + I'^{\,2}\;r_a + \text{stray power}}$$

As explained in a previous article, the *stray power* losses are made up of friction, eddy current and hysteresis losses. The stray power can be determined only by testing the dynamo.

In the *series-wound dynamo* there is but one circuit (see Fig. 40), and therefore but one current, *I*, the value of which

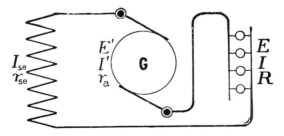

Fig. 40.   Diagram of Series-Wound Dynamo.

depends upon the value of the total voltage E' and the total resistance, which is made up of r_a, r_{se} and R.   The series field is composed of a comparatively few turns of heavy copper conductor and is connected in series with the armature and the external circuit.

The current may be expressed as

$$I' = I_{se} = I$$

$$I' = \frac{E'}{R + r_a + r_{se}}$$

$$I = \frac{E + (E' - E)}{R + (r_a + r_{se})} = \frac{E}{R} = I'$$

$$I_{se} = \frac{E' - E}{r_{se} + r_a} = I = I'$$

The total voltage may be expressed as

$$E' = E + I (r_a + r_{se}) = E \left\{ 1 + \frac{r_a + r_{se}}{R} \right\}$$

The electrical efficiency is

$$\eta_e = \frac{\text{useful energy}}{\text{total energy}} = \frac{E I}{E' I'} = \frac{E}{E'}$$

or in terms of resistances it may be expressed as

$$\eta_e = \frac{I^2 R}{I'^2 (R + r_a + r_{se})} = \frac{R}{R + r_a + r_{se}}$$

From the equation $I' = \dfrac{E'}{R + r_a + r_{se}}$ it is seen that an increase in the outside resistance diminishes the current in the field coils, thus diminishing the magnetic flux. As the constancy of the magnetic flux depends upon the constancy of the current value these series-wound dynamos are best adapted to give a constant current, and are used mostly for running arc lamps on series circuits.

From the equation $E' = E + I (r_a + r_{se})$ it is seen that the total current in the armature is the same as the current in the outside circuit, also that the total voltage is the external voltage plus the lost voltage in the armature and series coils due to the passage of the current.

The **Compound-Wound Dynamo** (see diagrams Figs. 9 and 10) may be considered as a shunt dynamo to which some series windings have been added to compensate for the fall in voltage at the brushes due to the demagnetizing effects of the armature, lost volts in armature, and to compensate for line losses as the load increases. The shunt circuit may either be connected to the brushes, in which case the machine is called the **Ordinary or Short Shunt Compound Dynamo** (see Fig. 9), or the shunt circuit may be connected to the terminals of the outside circuit in which case the machine is called a **Long Shunt Compound Dynamo** (see Fig. 10). Using the symbols as marked on these diagrams the equations for the ordinary or short shunt compound dynamo will be deduced. The total current is

$$I' = I + I_{sh} = I_{se} + I_{sh} = I + \frac{E + Ir_{se}}{r_{sh}}$$

It may also be written (see similar equation on page 66)

$$I' = \frac{E'}{r_a + \dfrac{(R + r_{se}) \times r_{sh}}{R + r_{se} + r_{sh}}}$$

or again it may be written

$$I' = I \left\{ 1 + \frac{R + r_{se}}{r_{sh}} \right\} = I \times \frac{r_{sh} + r_{se} + R}{r_{sh}}$$

The value of the shunt current may be written

$$I_{sh} = \frac{E' -- I'r_a}{r_{sh}} = \frac{E + Ir_{se}}{r_{sh}} = I \times \frac{r_{se} + R}{r_{sh}}$$

The value of the total E. M. F. generated may be written

$$E' = E + I' r_a + I r_{se}$$
$$= E + \left\{ I + \frac{E + I r_{se}}{r_{sh}} \right\} r_a + I r_{se}$$

The value of the total E. M. F. generated may also be expressed in terms of the useful voltage and the various resistances. It is equal to the sum of four lost voltages. The first of these voltages is E, the useful voltage. The next is the voltage lost in the series coil and is equal to $\dfrac{r_{se}}{R} \times E$.

Then the total voltage at the brushes is

$$\boldsymbol{E} + \frac{r_{se}}{\boldsymbol{R}} \, \boldsymbol{E} \quad \text{or} \quad \boldsymbol{E} \left\{ 1 + \frac{r_{se}}{\boldsymbol{R}} \right\} \quad \text{or} \quad \boldsymbol{E} \left\{ \frac{R + r_{se}}{\boldsymbol{R}} \right\}$$

The lost voltage in the armature due to that part of the current that flows through the series and outside circuit is to the voltage of the series and outside circuit as the resistance of the armature is to the resistance of the series and outside circuit.

$$\text{Lost volts in armature due to outside current} : E \left\} \frac{R + r_{se}}{R} \right\} :: r_a : r_{se} + R$$

$$\text{Lost volts in armature due to outside current} = E \left\} \frac{r_a (R + r_{se})}{R \times (r_{se} + R)} \right\}$$

Similarly the lost voltage in the armature due to the current flowing in shunt field is to the voltage of the shunt field as the resistance of the armature is to the resistance of the shunt field.

$$\text{Lost volts in armature due to shunt current} : E \left\} \frac{R + r_{se}}{R} \right\} :: r_a : r_{sh}$$

$$\text{Lost volts in armature due to shunt current} = \frac{E (R + r_{se}) r_a}{R \times r_{sh}}$$

Therefore, the total voltage generated being equal to the sum of these four expressions it may be written

$$E' = E + \frac{E r_{se}}{R} + \frac{E (R + r_{se}) r_a}{R (r_{se} + R)} + \frac{r_a (R + r_{se}) E}{r_{sh} \times R}$$

As the third term when simplified has the same denominator as the second term the two may be combined, and as all four terms in the right hand side of the equation contain E it may be taken outside the parenthesis ; then,

$$E' = E \left\{ 1 + \frac{r_{se} + r_a}{R} + \frac{r_a (R + r_{se})}{R \times r_{sh}} \right\}$$

The electrical efficiency of the ordinary compound dynamo is equal to the useful energy divided by the sum of the useful energy in the outside circuit, plus energy lost in series coils, plus energy lost in shunt coils, plus energy lost in armature.

$$n_e = \frac{E I}{E' I'} = \frac{I^2 R}{I^2 R + I^2 r_{se} + I_{sh}^2 r_{sh} + I'^2 r_a}$$

$$= \frac{I^2 R}{\dfrac{I^2 (r_{se} + r_{sh} + R)^2 r_a}{r_{sh}^2} + \dfrac{I^2 (r_{se} + R)^2}{r_{sh}} + I^2 R + I^2 r_{se}}$$

$$= \frac{1}{\dfrac{(r_{se} + r_{sh} + R)^2 r_a}{r_{sh}^2 R} + \dfrac{(r_{se} + R)^2}{r_{sh} R} + 1 + \dfrac{r_{se}}{R}}$$

For the **Long Shunt Compound-Wound Dynamo** (see Fig. 10) the equations may be written in a similar manner. The values of the total current generated will be

$$I' = I_{se} = I + I_{sh} = I + \frac{E}{r_{sh}} = I \frac{r_{sh} + R}{r_{sh}}$$

$$I_{sh} = \frac{E}{r_{sh}} = I \frac{R}{r_{sh}}$$

The total voltage generated in the armature is

$$E' = E + I'(r_a + r_{se}) = E + \left\{ I + \frac{E}{r_{sh}} \right\}(r_a + r_{se})$$

$$= E \left\{ 1 + \frac{R + r_{sh}}{R \, r_{sh}} \right\} (r_a + r_{se})$$

The electrical efficiency of the long shunt compound dynamo is

$$n_e = \frac{I^2 R}{I'^2 (r_a + r_{se}) + I_{sh}^2 r_{sh} + I^2 R}$$

$$= \frac{I^2 R}{\left\{ I^2 + \frac{2 R I^2}{r_{sh}} + \frac{R^2 I^2}{r_{sh}^2} \right\} (r_a + r_{se}) + \frac{I^2 R^2}{r_{sh}} + I^2 R}$$

$$= \frac{1}{\frac{r_a + r_{se}}{R} + 2 \left\{ \frac{r_a + r_{se}}{r_{sh}} \right\} + \frac{R (r_a + r_{se})}{r_{sh}^2} + \frac{R}{r_{sh}} + 1}$$

NOTE.

DESIGN OF DYNAMOS.

The complete calculations for the design of a dynamo are very long and involved. Besides calling for considerable judgment on the part of the designer, they require the use of a number of constants whose values must be known from previous experience or must be determined by testing the machine itself after completion. The most difficult part of the calculation is the assignment of the proper values to these constants.

In the following pages are given the electrical and magnetic details for the design of a 50-kilowatt shunt dynamo. The method is taken from Mr. A. E. Wiener's "Dynamo Electric Machines," and the constants from tables in that work, to which the reader is referred for further details. Though in general a 50-kilowatt machine should be made multipolar, the bipolar type is here chosen. The method is better illustrated and the calculations are simpler. For the multipolar machine the principle is the same, but the details differ somewhat.

The dynamo is a machine for converting mechanical into electrical energy, and must therefore be designed from the mechanical point of view as well as the electrical. Armature, shaft, bearings and pulley are therefore subject to the ordinary rules for machine design; and experience indicates that very conservative design is necessary in the parts subject to mechanical stresses. The following section, however, is limited to the electrical and magnetic parts of the design.

Calculation of a Bipolar, Single Magnetic Circuit, Smooth=Drum, High=Speed Shunt Dynamo.

50 K.W. Upright Horseshoe Type. Wrought=Iron Cores and Yoke, Cast=Iron Polepieces.
250 Volts. 200 Amperes. 1050 Revs. per Min.

(a) Calculation of Armature.

1. Length of Armature Conductor.

An electromotive force of one volt is generated by a conductor that cuts 100,000,000 C. G. S. lines of force per second. As the English system of units is still the standard in this country, one foot will be taken as the unit length of conductor, one foot per second the unit linear velocity, and one magnetic line of force per square inch as the unit field strength. Then every foot (12 in.) of inductor, moving at the rate of one foot (12 in.) per second in a field having one magnetic line of force per square inch will generate an E. M. F. of

$$\frac{12 \times 12 \times 1}{100,000,000} = \frac{144}{10^8} \text{ volt.}$$

As the armature of an ordinary bipolar dynamo has two parallel conductors each generating the same E. M. F.; and as these conductors are in parallel, two feet of conductor will be used in generating 144×10^{-8} volt or in other words each foot of the total length of conductors will generate only 72×10^{-8} volt if moving at unit velocity in unit field.

As this theoretical value of "unit armature induction" assumes that there is a magnetic field entirely surrounding the armature, it will have to be modified so as to take into account the fact that the fringe of magnetic lines from the pole pieces only partially surround the armature. For a 50 K.W. bipolar dynamo with smooth-drum armature the polar arc (see Figure 1) may be taken for the present as about 124°. It is found from actual practice that for a machine having a polar arc of about 124°, the unit armature induction will not be the theoretical value of 72×10^{-8} but will be about

$$e = 61 \times 10^{-8} \text{ volts.}$$

The "specific armature induction," $i.\ e.$, the induction per unit length of conductor moving at velocity v_c, in a magnetic field of strength, $\mathcal{H}''$, will be

$$e' = e \times v_c \times \mathcal{H}'' \text{ volts.}$$

where $e' =$ specific induction of active armature conductor, in volts per foot of conductor;

$e =$ unit armature induction per pair of armature circuits in volts per foot of conductor;

$v_c =$ conductor-velocity, or cutting speed, in feet per second;

$\mathcal{H}'' =$ field density, in magnetic lines of force per square inch.

It is customary to take the conductor velocity v_c, as about 50 feet per second for a 50 K.W. bipolar dynamo having a drum armature; also to take the field density $\mathcal{H}''$ as about 22,000 if the machine has cast iron pole pieces. Therefore the value of e' may be written,

$$e' = \frac{61 \times 50 \times 22,000}{100,000,000} = \frac{671}{1,000}$$

Knowing the specific armature induction, e', the voltage induced by one foot of conductor, and knowing the voltage, E', that the armature is required to induce, one may easily find the total length of *active* wire, L_a, of the armature.

$$L_a = \frac{E'}{e'} = \frac{E' \times 10^3}{671}$$

where $L_a =$ total length of active conductor (on whole circumference opposite pole pieces);

$E' =$ total E. M. F. to be generated in armature; $i.\ e.$, volt output plus additional volts to be allowed for drop due to internal resistance.

For a dynamo of 50 K.W. capacity it is necessary to add about 6% to the value of E, the voltage wanted by the external circuit, in order to get E', the total E. M. F. to be generated in the armature. If E is 250, E' is 106% of 250 or 265, and

$$L_a = \frac{265 \times 10^3}{671} = \mathbf{395\ feet} \text{ of } active \text{ conductor.}$$

The part of the conductor passing over the ends of the armature core is not active in cutting lines of magnetic force and so this

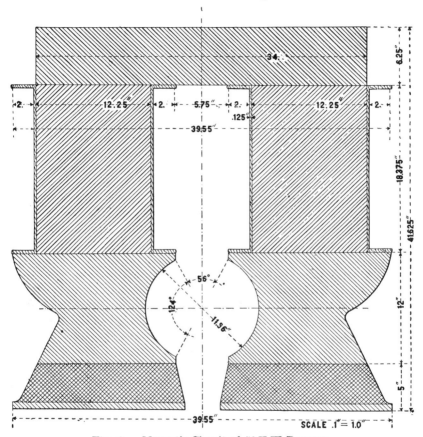

Fig. 1.   Magnetic Circuit of 50 K.W. Dynamo.

length of conductor will have to be increased when the dimensions of the iron core of the armature are known.

2.   Sectional Area of Armature Conductor, and Selection of Wire.

The cross section of wire to be chosen should be large enough so as not to be unduly heated by the current it has to carry.   It is well to allow about 600 circular mils of copper conductor to 1 ampere of current, which is a current density of about 2,100 amperes per square inch of copper. Therefore to get the number of circular mils of copper to be provided multiply 600 by the current to be generated. There is to be furnished to the outside circuit 200 amperes. The armature will have to furnish this plus the current that goes through the shunt field. This latter value is so small however that it may be omitted with only trifling error or one may choose slightly larger wire for the armature than is needed for the 200 amperes. The cross section of conductor required therefore will be

200 × 600 circular mils or 120,000 circular mils.

As there are two conductors to carry this current each conductor should have 60,000 circular mils. By reference to the B. & S. wire table on page 28 it will be seen that No. 2 wire has a cross section of 66,371 circular mils.

As No. 2 wire is too stiff to wind on the armature, however,

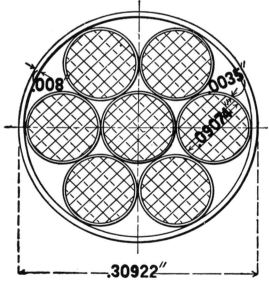

Fig. 2. Armature Conductor.

a cable made up of seven strands of No. 11 wire, having a copper cross section of 7 × 8,234 circular mils or a total of 57,638 circular mils is used. This will be a little higher current density than was calculated upon at first but is not at all excessive. The diameter of No. 11 B. & S. wire is .09074 inch and a single cotton insulation of .007 inch is put on each strand and the seven strands are covered with a double cotton insulation of .016 inch. This will give the cable composed of seven strands of No. 11 B. & S. insulated wire a diameter (see Fig. 2) of

$$\delta'_a = 3 \times (.09074 \text{ inch} + .007 \text{ inch}) + .016 \text{ inch}$$
$$= \mathbf{.30922 \text{ inch.}}$$

3. Diameter of Armature Core.

If the speed N is 1,050 revolutions per minute or $\dfrac{1050}{60}$ per second, and d'_a is the mean diameter of the armature winding in inches, or $\dfrac{d'_a}{12}$ in feet, then the cutting speed of the conductor in feet per second will be represented by,

$$v_c = \frac{d'_a \times \pi}{12} \times \frac{N}{60}$$

or

$$d'_a = \frac{12 \times 60}{\pi} \times \frac{v_c}{N} = \frac{12 \times 60 \times 50}{\pi \times 1050} = \mathbf{10.91''.}$$

It is found from actual practice that the ratio of the diameter of the armature, d_a, to the mean diameter of the armature

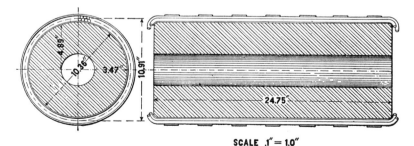

SCALE .1" = 1.0"

Fig. 3.   Details of Armature.

winding, d'_a, is about .95 for drum armatures between eight and ten inches in diameter. Therefore (see Fig. 3)

$$d_a = .95 \text{ of } 10.91 \text{ inches, or } \mathbf{10.36 \text{ inches.}}$$

4. Length of Armature Core.

Knowing the number of conductors that can be laid on the surface of the armature core, and the available depth for the windings, one may easily determine the number of conductors that can be wound on the armature; and knowing this and the length of active armature conductor, the length of the armature core may be determined.   For a machine generating less than 300 volts and having a drum armature of the size of the one under consideration about 8% of the surface of the armature core is given up to division strips or driving horns.

$$\text{Then } n_{\text{w}} = \frac{.92 \times d_{\text{a}} \times \pi}{\delta_{\text{a}}'}$$

$$= \frac{.92 \times 10.36 \times \pi}{.30922} = 97 \text{ or } 96.$$

where n_{w} = the number of armature wires per layer;
d_{a} = the diameter of the armature core, in inches;
δ'_{a} = the width of insulated armature conductor, in inches;
$.92$ = the portion of the surface to be occupied by the conductors.

For a drum armature of the size of the one under consideration the height of the windings should be about .55 of an inch and of this space about .06 inch will be taken up by the insulation between windings and the armature, and .05 inch by binding wires.   This leaves a net height of about .44 of an inch for the conductor.   Then the number of layers of conductors will be

$$n_1 = \frac{h'_{\text{a}}}{\delta''_{\text{a}}} = \frac{.44}{.30922} = 1\frac{1}{2} \text{ or } 2$$

where n_1 = number layers of armature wire;
h'_{a} = net height of winding space, in inches;
δ''_{a} = height of insulated conductor, in inches.

By dividing the length of active armature conductor by the number of conductors we obtain the length of one active conductor which is the length of the armature core (see Fig. 3).

$$l_{\text{a}} = \frac{12 \times L_{\text{a}}}{n_{\text{w}} \times n_1} = \frac{12 \times 395}{96 \times 2} = 24\frac{3}{4} \text{ inches approx.}$$

where l_{a} = length of armature core parallel to pole faces, in inches;
L_{a} = length of active armature conductor in feet;
n_{w} = number of wires per layer;
n_1 = number of layers of wire on armature.

5. Arrangement of Armature Windings.

For machines under 300 volts it is customary to have from 40 to 60 commutator segments. With this number of segments the voltage between two consecutive ones is low, and the pulsating current is within a fraction of one per cent of being a steady current. The number of commutator divisions, n_c, will be found by multiplying the number of wires per layer, n_w, by the number of layers, n_l, and dividing the product by some even number that gives a quotient which is between 40 and 60.

$$n_c = n_w \times n_l \div 4 = 48.$$

The number of convolutions per commutator segment, n_a will be

$$n_a = \frac{n_w \times n_l}{2 \times n_c} = \frac{96 \times 2}{2 \times 48} = 2$$

since it takes two conductors to make one turn. Therefore to sum up we have **48 coils,** each consisting of **2 turns** of a cable having **7 No. 11 B. & S. wires.**

6. Total Length of Armature Conductor, Weight and Resistance.

In order to connect the ends of the active conductors, turn the corners, etc., for this drum armature the total length of armature conductor will need to be about $1\frac{3}{4}$ times the active conductor.

$$L_t = 1.75 \times L_a = 1.75 \times 395 \text{ feet} = \textbf{691 feet.}$$

The weight of the conductor will be as follows: — A copper wire .001 in diameter weighs .00000303 pounds per foot of length. Therefore the weight of the total length of the copper in the armature conductor would be

$$wt_a = k_5 \times L_t \times \delta_a^2 \times .00000303$$
$$= 1.03 \times 691 \times 57638 \times .00000303 = \textbf{124 pounds}$$

where wt_a = weight of bare armature winding in pounds ;

k_5 = ratio between weights of the insulated wire and bare wire ;

L_t = total length of armature conductor in feet ;

δ_a^2 = area of conductor in circular mils.

The **resistance** of the armature will be as follows : — The total length of armature wire 691 feet is arranged in two parallel circuits $345\frac{1}{2}$ feet long. These two paths are each composed of seven No. 11 B. & S. wires.

The resistance of 1,000 feet of No. 11 B. & S. wire is 1.311 ohms.

Therefore the resistance of the armature is,

$$R_a = \frac{1}{2 \times 7} \times \frac{345.5}{1,000} \times 1.311 \text{ ohms} = .0324 \text{ ohm.}$$

7. Radial Depth of Armature Core, Minimum and Maximum Cross Section, and Average Magnetic Density of Armature Core.

The following formula determines the proper size for the necessary strength of the armature shaft where it passes through the iron core of the armature.

$$d_c = k_9 \times \sqrt[4]{\frac{P'}{N}} = 1.3 \times \sqrt[4]{\frac{50,000}{1,050}} = 3.42 \text{ inches.}$$

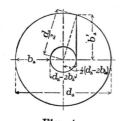

Fig. 4.

where d_c = diameter of armature shaft at core in inches ;

P' = capacity of machine in watts;

N = speed, in revolutions per minute ;

k_9 = constant depending upon capacity of machine.

Therefore the breadth of armature cross section, or radial depth of armature core is

$$b_a = \frac{1}{2}\,(d_a - d_c) = \frac{10.36 - 3.42}{2} = 3.47 \text{ inches.}$$

where b_a = radial depth of armature core, in inches ;

d_a = diameter of armature core, in inches ;

d_c = diameter of core section of armature shaft.

The maximum depth of armature core (see Fig. 4) is,

$$b'_a = \sqrt{\frac{d_a^2}{4} - \frac{(d_a - 2\,b_a)^2}{4}}$$

$$= \sqrt{\frac{(10.36)^2}{4} - \frac{(3.42)^2}{4}} = 4.89 \text{ inches.}$$

The cross section of the magnetic field in the armature is equal to

$$S''_{a1} = 2 \times l_a \times b_a \times k_2$$
$$= 2 \times 24.75 \times 3.47 \times .95 = \textbf{163 square inches.}$$
$$S''_{a2} = 2 \times l_a \times b'_a \times k_2$$
$$= 2 \times 24.75 \times 4.89 \times .95 = \textbf{230 square inches.}$$

where S''_{a1} = minimum cross section of armature core, in square inches;

S''_{a2} = maximum cross section of armature core, in square inches;

l_a = length of armature core, in inches;

b'_a = radial depth of armature core, in inches;

k_2 = ratio of net iron section to total cross section of armature core.

The useful magnetic flux in the armature may be obtained as follows : The

$$\text{E. M. F.} = \frac{\text{Number of C. G. S. lines cut per second}}{10^8}$$

$$E' = \frac{N \times N_c \times \Phi}{60 \times 10^8} \text{ volts}$$

where E' = total E. M. F. induced in armature ;

N = number of armature revolutions per minute ;

N_c = total number of conductors all around pole facing surface of armature;

Φ = total number of useful magnetic lines, in webers

or $\quad \Phi = \dfrac{60 \times 10^8 \times E'}{N \times N_c}$

$$= \frac{60 \times 10^8 \times 265}{1050 \times 192} = \textbf{7,886,905 webers.}$$

Therefore the density of magnetic lines per square inch at the minimum cross section of the armature core is

$$\mathfrak{B}''_{a1} = \frac{7,886,905}{163} = \textbf{48,386 lines} \text{ per square inch.}$$

and the density at the maximum cross section is

$$\mathfrak{B}''_{a2} = \frac{7,886,905}{230} = \textbf{34,291 lines} \text{ per square inch.}$$

Next we wish to find the number of ampere-turns required to furnish this induction. The law for the magnetism in a magnetic circuit is exactly similar to the law for current in an electric circuit.

$$\text{Current (or Electric Flux)} = \frac{\text{Electromotive Force}}{\text{Resistance}}.$$

$$\text{Magnetic Flux} = \frac{\text{Magnetomotive Force}}{\text{Reluctance}}.$$

Therefore the

$$\text{Magnetomotive Force} = \text{Magnetic Flux} \times \text{Reluctance}.$$

As the resistance of an electric circuit can be expressed by the specific resistance or resistivity of the material multiplied by the length of circuit and divided by the cross section, so the reluctance of a magnetic circuit can be expressed by the specific reluctance or reluctivity of the material multiplied by the length of the circuit and divided by the cross section. Therefore the

$$\text{Reluctance} = \text{Reluctivity} \times \frac{\text{Length}}{\text{Area}}.$$

As the conductivity of an electric circuit is the reciprocal of the resistivity, so the permeability of a magnetic circuit is the reciprocal of the reluctivity. Therefore the

$$\text{Reluctance} = \frac{\text{Length}}{\text{Permeability} \times \text{Area}}.$$

or

$$\text{Magnetomotive Force} = \frac{\text{Magnetic Flux} \times \text{Length}}{\text{Permeability} \times \text{Area}}.$$

and since the magnetic flux divided by the area is the magnetic density the formula is simplified by writing

$$\text{Magnetomotive Force} = \frac{\text{Magnetic Density} \times \text{Length}}{\text{Permeability}}.$$

The Unit of Field Density, or the magnetic density caused by a unit pole, is 1 line of magnetic force per square centimeter of field area and is termed 1 gauss.

A single Line of Force, or the Unit of Magnetic Flux, is that amount of magnetism that passes through every square centimeter of cross section of a magnetic field whose density is unity, and is termed 1 weber.

The unit magnetic pole, or the pole of unit strength is that which repels an equal pole at unit distance with unit force. The lines of force are straight lines from the center of the sphere to the surface, there being one line to each square centimeter area on the surface. As the surface of a sphere having a radius of 1 centimeter has an area of 4π square centimeters, it follows that from a pole of unit strength there is a magnetic flux of 4π C. G. S. lines of magnetic force or 4π webers or 12.5664 webers.

One absolute C. G. S. unit of current, which is 10 times as large as the ampere, or ten amperes, flowing in a wire which is bent into a circle of one centimeter radius gives a C. G. S. unit magnetic pole at the center of curvature or 12.5664 webers. One practical unit or 1 ampere would cause one tenth as many webers or 1.25664 webers.

A long solenoid having a cross section of one square centimeter having 1 ampere ($\frac{1}{10}$ of the C. G. S. unit current) flowing per unit length of coil, has poles of $\frac{1}{10}$ unit strength which causes a magnetic flux of $\frac{1}{10} \times 4\pi$ webers.

The density of the magnetic circuit is $\frac{4\pi}{10}$ webers per square centimeter or $\frac{4\pi}{10}$ gausses.

The reluctance of unit length of the solenoid of one square centimeter cross section for air is unity or 1 oersted. The magnetomotive force is the product of the magnetic density, the reluctance and the length, and is measured in gilberts. Therefore the magnetomotive force required to produce a magnetic density of $\frac{4\pi}{10}$ gausses in a column of air one centimeter long and having a cross section of one square centimeter thus having a reluctance of 1 oersted is

$$\text{M. M. F.} = \frac{4\pi}{10} \times 1 \times 1 = \frac{4\pi}{10} \text{ gilberts.}$$

The magnetomotive force of $\frac{4\pi}{10}$ gilberts being produced by one ampere-turn, it follows that the

$$\text{Number of Ampere-turns} = \frac{10}{4\pi} \times \text{ Number of Gilberts.}$$

The

Magnetizing Force = Specific Magnetizing Force × Length,
or the

Number of Ampere-turns = Ampere-turns per unit of Length × Length,

or $a\,t = f\,(\text{\ss}'') \times l$

where $a\,t$ = Ampere-turns required to magnetize a portion of a magnetic circuit;

$\text{\ss}''$ = density of the magnetic circuit per square inch;

$f\,(\text{\ss}'')$ = Specific magnetizing force, in ampere-turns per inch of length for the particular material and density employed; (this value of the magnetizing force must be taken from some table or induction curve as shown in Fig. 5, or found by experiment for the particular piece of iron to be used);

l = length of magnetic circuit of the material in inches.

To get the number of ampere-turns required to overcome the reluctance of the armature it is necessary to modify the above formula a little, as the value of $\text{\ss}''$ is not constant in all parts of the core.

$$f\,(\text{\ss}''_a) = \frac{1}{2}\left\{ f\,(\text{\ss}''_{a1}) + f\,(\text{\ss}''_{a2}) \right\}$$

$$= \frac{f\,(48,386) + f\,(34,291)}{2} = \frac{9.2 + 6.4}{2} = 7.8$$

It takes as seen from Fig. 5 about 9.2 ampere-turns to mag netize 1 inch of wrought iron to a density of 48,386 lines, and 6.4 ampere-turns to magnetize it to a density of 34,291 lines. Also 7.8 ampere-turns correspond to a density of **41,500 lines.**

8.   Energy Losses in Armature, and Temperature Increase.

The energy lost in the armature due to the current in the armature conductors is ·

$$P_a = 1.2 \times (k_6 \times I)^2 \times r_a$$
$$= 1.2 \times (1.03 \times 200)^2 \times .0324 = \textbf{1650 watts}$$

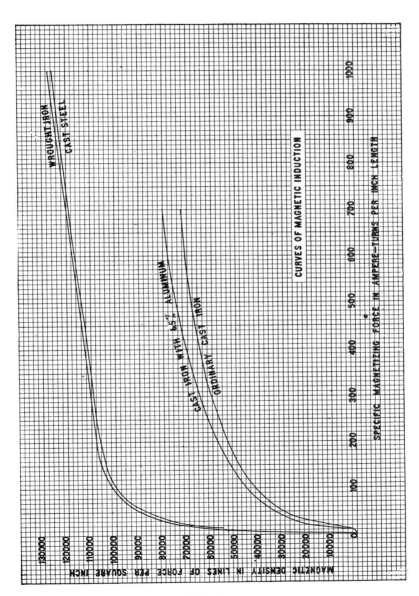

Fig. 5.

where P_a = energy dissipated in armature winding, in watts;

r_a = resistance of armature winding cold, in ohms ;

1.2 = ratio of resistance of armature winding hot, to resistance cold ;

I = output of machine in amperes ;

1.03 = ratio of total current generated to output.

The resistance of copper wire at 150° F. is about 1.2 times that at 60° F. The energy in the shunt coil of a 50 K.W. compound dynamo is about 3 % of the output of the machine.

The loss in the armature core due to hysteresis is proportional to the 1.6th power of the magnetic density, directly proportional to the number of magnetic reversals, and directly proportional to the mass of the iron. Expressed in C. G. S. absolute units the energy consumed by hysteresis is,

$$P_h' = n_1 \times \mathcal{B}_a{}^{1\cdot 6} \times N_1 \times M_1 ,$$
$$P_h' = n_1 \times \mathcal{B}_a{}^{1\cdot 6} \times N_1 \times M_1' ,$$

where P'_h = energy lost due to hysteresis, in ergs;

n_1 = constant depending on magnetic hardness of material Hysteresis Resistance ;

$\mathcal{B}_a$ = density of lines per square centimeter of iron ;

N_1 = frequency, or number of complete cyles of 2 reversals each, per second;

M_1' = mass of iron in cubic centimeters.

For soft sheet iron discs n_1 may be taken as .0035. In order to get the value of P_h' in practical units we must change the above equation. One watt equals 10^7 ergs. Instead of $\mathcal{B}_a$ use the density $\mathcal{B}_a''$ in lines per square inch.

$$N_1 = \frac{N}{60} = \frac{1050}{60} = 17\tfrac{1}{2}.$$

One cubic foot equals 28,316 cubic centimeters. The mass of iron in the armature in cubic feet is

$$M = \frac{d_a{}''' \times \pi \times b_a \times l_a \times k_2}{1,728}$$

$$\frac{(10.36 - 3.47) \times \pi \times 3.47 \times 24.75 \times .95}{1728} = 1.02 \text{ cubic ft.,}$$

where M = net mass of iron, in cubic feet;

d_a''' = mean diameter of armature core, in inches,

= $d_a - b_a$, (see Fig. 4);

l_a = length of armature core, in inches;

b_a = radial depth of armature core, in inches;

k_2 = ratio of net iron section to total iron section.

There are 1,728 cubic inches in 1 cubic foot. Expressing the value of the energy lost by hysteresis in practical units we have,

$$P_h = 10^{-7} \times .0035 \times \left\{ \frac{\mathfrak{B}_a''}{6.45} \right\}^{1.6} \times 28,316 \times N_1 \times M,$$

where P_h = energy lost by hysteresis, in watts;

$\mathfrak{B}_a''$ = density, in lines per square inch, corresponding to average magnetizing force required for armature core, (1 square inch = 6.45 square centimeters);

N_1 = frequency, in cycles per second;

M = net mass of iron in armature, in cubic feet.

By reducing the above expression to simple form, we have

$$P_h = 5 \times 10^{-7} \times \mathfrak{B}_a''^{1.6} \times N_1 \times M,$$
$$= 5 \times 10^{-7} \times (41,500)^{1.6} \times 17\frac{1}{2} \times 1.02 = 219 \text{ watts.}$$

Eddy Currents.

It has been explained under *Electromagnetic Induction* (see "Theory of Dynamo Electric Machinery") by what means and methods electromotive forces, and consequently electrical currents are set up in closed conductors. It is clear that any mass of metal moving in a field is a closed conductor. A loop of wire may have currents generated in it quite as easily if it be made part of a metallic disc as if it were still a loop. Since the lines of force can only cut a solid piece of metal once, the electromotive force that will be generated, unlike that in a coil of many turns cut by the same field, will be very small. But if the piece of metal be large the resistance will be very small so that the current which is induced may reach considerable strength and cause much heating. This action is largely prevented from taking place in the armature cores of dynamos by building them up of thin sheets of iron. The sheets are insulated from each other and are placed in such a direction as to cut across the path which would be followed by the induced current. This method of construction is termed lamination, and it is necessary to build all iron or metallic parts which are not intended to act as conductors in this way, if they are likely to be subjected to fields of varying strength. This is the case in alternating current machinery and in the moving parts of direct current machinery. Sometimes even conductors, if they are very large, have local currents generated in them which are of course undesirable because they are a source of waste and heat, and it becomes necessary to laminate them. In this case the lamination would be parallel to the length of the conductor. Heavy conductors are usually laminated or stranded any way for greater ease in construction. Eddy currents are often called Foucault currents from Foucault who first called attention to their existence.

The energy lost by eddy currents is found to be proportional to the square of the magnetic density, to the square of the frequency, and to the mass. The equation for the lost energy due to eddy currents is,

$$P_e' = \epsilon' \times \mathcal{B}_a{}^2 \times N_1{}^2 \times M'_1 ;$$

where P_e' = lost energy due to eddy currents, in ergs ;

     $\mathcal{B}_a$ = density of lines of force, per square centimeter of iron ;

     N_1 = frequency, in cycles per second ;

     M_1' = mass of iron, in cubic centimeters ;

     ϵ' = eddy current constant, depending upon the thickness and the specific electric conductivity of the material ;

$$\epsilon' = \frac{\pi^2}{6} \times \delta^2 \times \gamma \times 10^{-9}$$

$$= 1.645 \times \delta^2 \times \gamma \times 10^{-9}$$

where δ = thickness of sheet iron, in centimeters ;

     γ = electrical conductivity, in mhos ;

     γ = 100,000 for iron.

By changing to practical units and simplifying the above equation,

$$P_e = 10^{-7} \times 1.645 \times (2.54\ \delta_i)^2 \times 10^{-4} \times \left\{ \frac{\mathfrak{B}_a''}{6.45} \right\}^2 \times N_1^2$$

$$\times\ 28,316 \times M = 7.22 \times 10^{-8} \times \delta_i^2 \times \mathfrak{B}_a''^2 \times N_1^2 \times M.$$

where δ_i = thickness of iron laminæ in armature, in fractions of an inch;

$\mathfrak{B}_a''$ = density, in lines per square inch, corresponding to average specific magnetizing force of armature core;

N_1 = frequency in cycles per second;

M = mass of iron, in cubic feet.

The terms 7.22, 10^{-8}, δ_i^2, and $\mathfrak{B}_a''^2$, may be multiplied together to form the eddy current factor ϵ, and then the formula for loss due to eddy currents becomes,

$$P_e = \epsilon \times N_1^2 \times M$$
$$= .0125 \times (17\tfrac{1}{2})^2 \times 1.02 = 4.\ \text{watts}.$$

For $\epsilon = .0125$ when the thickness of the sheet iron discs is .01″ and when the magnetic density is 41,500 lines.

The total energy loss in the armature is
$$P_A = P_a + P_h + P_e = 1650 + 219 + 4$$
$$= 1873\ \text{watts} = 2.5\ \text{h. p.}$$

Knowing the amount of electrical energy lost in the armature and dissipated as heat, and knowing the dimensions and speed of the armature it is possible, by using certain known constants that have been found by experience, to calculate quite exactly the rise of temperature in the armature and to calculate the hot resistance of the copper in the armature conductors. These details are somewhat complicated and will not be treated.

(b) Dimensions of Magnet Frame.

1. Total Magnetic Flux, and Sectional Area of Magnet Frame.

The total magnetic flux to be generated in a dynamo is the useful magnetic flux multiplied by a factor of magnetic leakage. Knowing the shape of the various parts of the magnetic circuit and the reluctance of the parts of the magnetic circuit and the surrounding air circuits, it is possible to determine the amount of magnetic leakage, or waste magnetism.

$$\Phi' = \lambda \times \Phi = 1.35 \times 7{,}886{,}905 = 10{,}647{,}322 \text{ webers}$$

where Φ' = total flux to be generated in dynamo, in lines of force ;

Φ = useful flux necessary to produce the required E. M. F.;

λ = factor of magnetic leakage, which for a 50 K.W. bipolar machine of the Edison type is found to be about 1.35.

The sectional area of the magnet frame is expressed by the formula,

$$S''_m = \frac{\Phi'}{\mathfrak{B}''_m} = \frac{10{,}647{,}322}{90{,}000} = \textbf{118 square inches}$$

where S''_m = cross section of wrought iron magnet core and of yoke, in square inches ;

Φ' = total flux, in webers ;

$\mathfrak{B}''_m$ = magnetic density of magnet frame, which for wrought iron magnet cores and yoke is taken as 90,000 lines per square inch.

The Practical Limit of Magnetization for cast iron pole pieces is taken as 50,000 lines per square inch.

$$S''_m = \frac{\Phi'}{\mathfrak{B}''_m} = \frac{10{,}647{,}322}{50{,}000} = \textbf{213 square inches.}$$

2. Sectional Area of Magnet Frame.

If the magnet core is in the form of a cylinder the diameter will be

$$d_m = \sqrt{\frac{4}{\pi} \times 118} = 12.25 \text{ inches.}$$

For a circular magnet core carrying about 10,000,000 lines of force the most economical ratio of length to diameter of magnet core is found to be about 1.5.   Hence

$$l_m = 1.5 \times d_m = 18.375 \text{ inches.}$$

For a machine having an armature between 10 and 11 inches in diameter the distance, c, between magnet cores is taken as about

$$c = 5\tfrac{3}{4} \text{ inches.}$$

Take the width of the wrought iron yoke a little greater than the diameter of the magnet cores so as to form a mechanical protection to them. Let this width be $18\frac{3}{4}$ inches. Then the height of the yoke would be

$$h_y = \frac{118}{18\frac{3}{4}} = 6\frac{1}{4} \text{ inches.}$$

The pole pieces are determined as follows : — The bore of the pole pieces is the sum of the diameter of the armature core, the winding, the insulation and binding, and the air gaps, which for a drum armature of this diameter is taken as about $.14''$.

Diam. of bore $= 10.36'' + 4 \times .30922'' + 2 \times .11'' + .14''$
$= 11.96''$.

In order to keep the corners of the pole pieces far enough apart so that there will not be too much leakage across, the distance is taken from 1.25 to 8 times the length of the two air gaps according to the size and type of dynamo.

$$l'_p = k_{11} \times (d_p - d_a) = 4. \times (11.86 - 10.36) = 6.00$$
inches.

where l'_p = distance between pole corners, in inches ;
d_p = diameter of bore of pole pieces, in inches ;
d_a = diameter of iron core of armature, in inches ;
k_{11} = 4. for drum armature of a 50 K.W. bipolar dynamo.

Let the length of the pole pieces equal the length of iron core in armature, or

$$l_p = 24\frac{3}{4}''.$$

Let the height of pole pieces equal the diameter of bore approx.

$$h_p = 12''.$$

The thickness at the center which carries only half the lines will be

$$\frac{213}{2 \times 24\frac{3}{4}} = 4.3 + \text{ or } 4\frac{3}{8}''.$$

In order to provide a non-magnetic base through which the magnetic lines will not leak from pole to pole a block of zinc about 5 inches thick should be placed under the dynamo.

(c) Calculation of Magnetizing Forces.

1. Air Gaps.

The length of the two air gaps is

$$l''_g = k_{12} \times (d_p - d_a) = 1.4 \times 1.5'' = 2.1''$$

where l''_g = length of path of magnetic lines across the two air
         gaps;

d_p = diameter of bore of pole pieces, in inches;

d_a = diameter of iron core of armature, in inches;

k_{12} = constant depending on the path of the lines of force
through the air gap. The lines of force pass through the gaps
obliquely owing to the distortion of the magnetic field and the
constant, k_{12}, grows greater as the velocity of the conductor, v_c,
and the density, $\mathcal{K}$, are greater. If the product of v_c and $\mathcal{K}$ is
above 2,000,000 the value of k_{12} is taken as 1.4.

The cross section of the magnetic field of the air gap is represented by,

$$S_f = d_f \times \frac{\pi}{2} \times \beta'_1 \times l_f$$

$$= 11.11'' \times \frac{\pi}{2} \times .84 \times 24\tfrac{3}{4} = 363 \text{ square inches}$$

where S_f = the area occupied by the effective conductors, in square
         inches ;

d_f = mean diameter of magnetic field in inches ;

     $= \tfrac{1}{2} (d_a + d_p)$;

l_f = breadth of magnetic field, in inches ;

β'_1 = ratio of effective field circumference, depending upon
         the percentage of polar embrace, here taken as .84.

The actual field density in air gaps is found by dividing the
value of the total useful flux in webers, by the area of the
magnetic field of air gap.

$$\mathcal{K}'' = \frac{\Phi}{S_f} = \frac{7,886,905}{363} = 21,727 \text{ lines per square inch.}$$

The number of ampere-turns required to produce this magnetic density in the air gaps is,

$$a\,t_g = \frac{10}{4\pi} \times \mathcal{K}'' \times \frac{l''_g}{2.54} = .3133 \times \mathcal{K}'' \times l''_g$$

$$= .3133 \times 21,727 \times 2.1 = \mathbf{14,295\ ampere\text{-}turns.}$$

2. Armature Core.

The average length of the magnetic paths in the armature is (see Fig. 6).

$$l''_a = d'''_a \times \pi \times \frac{90° + a}{360°} + b_a$$

$$= 6.89'' \times \pi \times \frac{90 + 28}{360} + 3.47'' = \textbf{10.57 inches.}$$

where l''_a = length of magnetic path in armature core, in inches;

d'''_a = mean diameter of armature core, in inches;

b_a = radial depth of armature core, in inches;

a = half angle between adjacent pole corners.

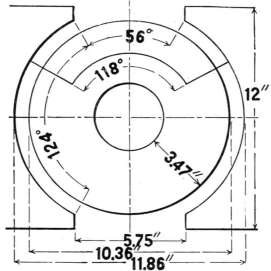

Fig. 6. Magnetic Details of Armature Core.

As previously determined the minimum cross section of the armature core is,

$$S''_{a1} = 163 \text{ square inches,}$$

and the maximum cross section is

$$S'''_{2a} = 230 \text{ square inches.}$$

The average specific magnetizing force to magnetize the armature core to the required density will be

$$f(\mathcal{B}''_a) = \frac{f(48{,}386) + f(34{,}291)}{2} = 7.8 \text{ ampere-turns.}$$

The total magnetizing force required to magnetize the armature is

$a\ t_a = 10.57 \times 7.8 = $ **82 ampere-turns.**

3. Wrought Iron Field Cores and Yoke.

The length of the magnetic path through the wrought iron cores and the yoke will be (see Fig. 1),

$l''_{w.i.} = 2 \times 18\frac{3}{4} + 6\frac{1}{4} + 18 = $ **61 inches.** (approx.)

The density of the field has already been decided upon as 90,000 lines per square inch. It has been determined by experiment that a specific density of 90,000 lines in wrought iron is produced by 50.7 ampere turns.

The total length of magnetic circuit through field cores and yoke will require

$a\ t_{w.i.} = 61 \times 50.7 = $ **3093 ampere-turns.**

4. Cast Iron Pole Pieces.

The average length of the magnetic path through the pole pieces may be taken as the average of the longest path and the shortest path. In the present case it is about

$l''_{c.i.} = 15$ inches.

The minimum cross section at the center of the pole piece which carries only half the lines has already been taken as

$S_{c.i.1} = 106.5$ square inches

which corresponds to a density of

$\mathfrak{B}''_{c.i.} = 50,000$ lines per square inch.

The maximum cross section of the magnetic circuit in pole pieces is the area of the pole face and is

$$S_{c.i.2} = 11.86 \times \pi \times \frac{124}{360} \times 24.75 = 318 \text{ square inches,}$$

which corresponds to a minimum density of

$$\mathfrak{B}''_{c.i.2} = \frac{7,886,905}{318} = 24802 \text{ magnetic lines.}$$

The average specific magnetizing force will be

$$f\ (\mathfrak{B}''_{c.i.}) = \tfrac{1}{2}\ [f\ (50,000) + f\ (24,802)] = \frac{160 + 38.6}{2}$$

$$= 99.3 \text{ ampere-turns,}$$

which corresponds to average density of **42,175 lines.**

Then the total number of ampere-turns required to magnetize pole pieces is

$$a\, t_{\text{c.i.}} = 15 \times 99.3 = \textbf{1490 ampere turns.}$$

5. Armature Reactions.

The demagnetizing and cross-magnetizing effects caused by the ampere-turns of the armature oppose the magnetizing effect of the field coils and have to be overcome by adding more ampere-turns to those required to overcome the reluctance of the circuit.

The number of ampere-turns required to balance the armature reaction is,

$$a\, t_{\text{r}} = k_{14} \times \frac{N_{\text{a}} \times I'}{2} \times \frac{k_{13} \times a}{180°}$$

$$= 1.71 \ \times \frac{96 \times 200}{2} \times \frac{28}{180} = \textbf{2554 ampere-turns}$$

where $a\, t_{\text{r}}$ = ampere-turns required to compensate for armature reactions;

k_{14} = being a constant depending upon the magnetic density in pole pieces;

N_{a} = total number of turns on armature;

I' = total armature current in amperes;

2 = number of armature circuits for current;

$k_{13} \times a$ = angle of brush lead, which is nearly equal to half the angle between two pole corners, for smooth drum armatures.

The total number of ampere-turns required on field coils of dynamo will be the sum of all the ampere-turns required for the various parts of the circuit.

$$AT = at_{\text{g}} + at_{\text{a}} + at_{\text{w.i.}} + at_{\text{c.i.}} + at_{\text{r}}$$
$$= 14{,}295 + 82 + 3093 + 1490 + 2554$$
$$= \textbf{21,514 ampere-turns.}$$

(d) Calculation of Magnet Winding.

The shunt winding should be calculated for a temperature increase of about 15°C above the normal temperature. (If desired the shunt coil may be calculated so as to give a higher voltage to the dynamo which can be decreased by a regulating resistance in series with the shunt coil. As this calculation is not required in order to show the electromagnetic theory of the dynamo it will not be discussed here. It requires a recalculation of the magnetic flux for the various parts of the magnetic circuit and a recalculation of the corresponding ampere-turns.)

1. Magnet Winding.

The mean length of one turn of wire on the field core is

$$l_{\text{T}} = k_{17} \times d_{\text{m}} = 3.66 \times 12\tfrac{1}{4}'' = 44.8''$$

where k_{17} = a constant, depending upon the size of the field core, giving the ratio of the length of a mean turn to the core diameter.

The specific length of magnet shunt wire in feet per ohm is given by the formula,

$$\lambda_{\text{sh}} = \frac{L_{\text{sh}}}{r_{\text{sh}}}$$

$$= \frac{AT}{E} \times \frac{l_{\text{T}}}{12} \times (1 + .004 \times \theta_{\text{m}})$$

$$= \frac{21{,}514}{250} \times \frac{44.8}{12} \times (1 + .004 \times 15)$$

$$= 340.5 \text{ feet per ohm.}$$

No. 15 B & S has 315 feet per ohm.

No. 14 B & S has 397 feet per ohm.

Use about No. 14 B & S wire.

The height of the winding space is given by the formula

$$h_{\text{m}} = \frac{l_{\text{T}}}{\pi} - d_{\text{m}} = \frac{44.8}{\pi} - 12.25 = 2 \text{ inches.}$$

The radiating surface of the magnet coils will be

$$S_{\text{M}} = (12.25 + 2 \times 2)\,\pi \times 2\,(18.375) = 1877 \text{ square inches.}$$

The energy absorbed in the magnet winding may be expressed by the formula,

$$P_{sh} = \frac{\theta_m}{75} \times S_M$$

$$= \frac{15}{75} \times 1877 = \textbf{375 watts.}$$

where P_{sh} = watts absorbed in field winding;

θ_m = rise of temperature in magnets, in degrees Centigrade ;

S_M = radiating surface of magnet coils.

$$I_{sh} = \frac{P_{sh}}{E} \quad \text{or } I_{sh}\, E = P_{sh}.$$

Therefore the number of shunt turns may be found by

$$N_{sh} = \frac{A\,T}{I_{sh}} = \frac{A\,T \times E}{P_{sh}}$$

$$= \frac{21514 \times 250}{375} = \textbf{14,343 shunt turns.}$$

The length of the shunt winding is

$$L_{sh} = \frac{14{,}343 \times 44.8}{12} = \textbf{53,547 feet.}$$

The resistance of the shunt wire will be

$$r_{sh} = \frac{53{,}547}{340.5} = \textbf{157 ohms}, \text{ at } 15.5^\circ \text{ C.}$$

The warm resistance at 30.5° C is

$$r_{sh} = 157 \times (1 + .004 \times 15) = \textbf{166 ohms.}$$

The shunt current at full load is

$$I_{sh} = \frac{250}{166} = \textbf{1.5 amperes.}$$

If desired it is a simple matter to get the weight of the wire used.

(e) Calculation of Efficiency.

The electrical efficiency of the dynamo is the ratio of the available energy to the available energy plus the energy lost due to armature resistance plus that due to field coil resistance.

$$\eta_e = \frac{P}{P + P_a + P_M}$$

$$= \frac{250 \times 200}{250 \times 200 + 1.06 \times 201.5^2 \times .0324 + 1.5^2 \times 166}$$

$= .96$ or **96% electrical efficiency.**

The commercial efficiency is the ratio of the output, to the sum of the output, wire loss in armature, wire loss in field, hysteresis loss, eddy current loss, and friction loss (which we will assume as 2,500 watts).

$$\eta_c = \frac{P}{P + P_a + P_M + P_h + P_e + P_o}$$

$$= \frac{50,000}{50,000 + 1,394 + 374 + 219 + 4 + 2,500}$$

$= .917$ or **91.7%, commercial efficiency.**

WIRE TABLE — PROPERTIES OF COPPER WIRE

No. B.&S. Gauge	Diameter "d" Mils (1 Mil=.001 in)	Area — Circular Mils d²	Area — 1000ths of an Inch d²×.7854	Wt. — Lbs per 1000 Ft	Wt. — Lbs per Mile	Wt. — Feet per Lb	Wt. — Ohms per Lb Annealed	Ohms per 1000 Ft Annealed	Ohms per 1000 Ft Hard-Drawn	Ohms per Mile Annealed	Ohms per Mile Hard-Drawn	Feet per Ohm Annealed	Tensile Annealed	Tensile Hard-Drawn	Elong. % Annealed	Elong. % Hard-Drawn	Safe Cur. Bright Wire Paneled	Safe Cur. Black Wire Free Air	Turns per Lin. In. Cotton Cov.
0000	460.000	211600.00	.166190	640.5	3381.4	1.561	.00007639	.04893	.050036	.25835	.26419	20440	5650	9975	45.0	5.00	146.0	346.0	1.6
000	409.642	167806.43	.131794	508.0	2682.2	1.969	.0001215	.06170	.063094	.32577	.33314	16210	4475	7900	45.0	5.00	127.0	292.0	1.6
00	364.796	133076.66	.104518	402.8	2126.8	2.482	.0001931	.07780	.079558	.41079	.42007	12850	3550	6250	45.0	4.00	110.0	247.0	1.8
0	324.861	105534.50	.082887	319.5	1686.9	3.130	.0003071	.09811	.10033	.51802	.52973	10600	2800	4950	45.0	4.00	95.0	209.0	2.8
1	289.296	83692.67	.065732	253.3	1337.2	3.947	.0004883	.1237	.12649	.65314	.66790	8083	2225	3925	42.0	4.00	83.0	177.0	3.2
2	257.626	66371.31	.052128	200.9	1060.6	4.977	.0007765	.1560	.15953	.82368	.84239	6410	1750	3125	39.0	3.00	72.0	148.0	3.4
3	229.407	52634.37	.041339	159.3	841.1	6.276	.001235	.1967	.20114	1.0386	1.0621	5084	1400	2450	36.0	3.00	63.0	127.0	3.9
4	204.307	41741.32	.032784	126.4	667.4	7.914	.001936	.2480	.25361	1.3094	1.3392	4031	1100	1950	36.0	3.00	54.0	108.0	4.5
5	181.941	33102.37	.025999	100.2	529.1	9.980	.003122	.3128	.31987	1.6516	1.6889	3197	875	1550	30.5	2.50	48.0	91.0	5.0
6	162.022	26251.37	.020618	79.46	419.6	12.58	.004963	.3944	.40332	2.0825	2.1295	2535	700	1225	28.0	2.00	42.0	78.0	5.6
7	144.428	20818.35	.016351	63.02	332.7	15.87	.007892	.4973	.50854	2.6258	2.6850	2011	550	975	28.0	2.00	37.0	66.0	6.2
8	128.490	16509.64	.012967	49.98	263.9	20.01	.01255	.6271	.64127	3.3111	3.3859	1595	425	775	24.0	1.75	32.0	56.0	7.0
9	114.434	13092.75	.010283	39.63	209.2	25.23	.01995	.7908	.80876	4.1753	4.2769	1265	325	600	21.5	1.75	28.2	48.0	7.5
10	101.897	10383.02	.0081548	31.43	165.9	31.82	.03173	.9972	1.0199	5.2657	5.3848	1003	275	475	19.0	1.50	24.9	41.0	8.5
11	90.743	8234.11	.0064656	24.93	131.6	40.12	.05045	1.257	1.2854	6.6369	6.7869	795.3	200	375	16.0	1.50	21.9	35.0	9.7
12	80.808	6529.95	.0051287	19.77	104.4	50.59	.08022	1.586	1.6218	8.3741	8.5633	630.7	175	300	16.0	1.25	19.3	30.0	11.2
13	71.962	5178.48	.0040672	15.68	82.79	63.79	.1276	1.999	2.0443	10.555	10.794	500.1	125	225	13.0	1.25	17.0	25.8	12.0
14	64.084	4106.72	.0032254	12.43	65.63	80.44	.2028	2.521	2.5779	13.311	13.612	396.6	100	200	11.0	1.25	15.0	22.2	13.0
15	57.069	3256.78	.0025579	9.858	52.05	101.4	.3225	3.179	3.2508	16.785	17.165	314.5	80	150	9.0	1.25	13.3	19.1	15.3
16	50.821	2582.74	.0020285	7.818	41.28	127.9	.5128	4.009	4.0996	21.168	21.646	249.4	60	125	8.0	1.25	11.8	16.5	16.7
17	45.257	2048.30	.0016087	6.200	32.74	161.3	.8153	5.055	5.1692	26.691	27.294	197.8	50	90	7.0	1.25	10.4	14.2	17.7
18	40.303	1624.30	.0012757	4.917	25.96	203.4	1.296	6.374	6.5180	33.655	34.416	156.9	40	70	6.0	1.20	9.2	12.3	19.5
19	35.890	1288.13	.0010117	3.899	20.59	256.5	2.061	8.038	8.2196	42.441	43.400	124.4	30	55	5.0	1.20	8.2	10.2	22.7
20	31.961	1022.53	.00080231	3.092	16.33	323.4	3.278	10.14	10.372	53.539	54.749	98.66	25	45	5.0	1.00	7.2	9.2	25.0
21	28.463	810.12	.00063626	2.452	12.95	407.8	5.212	12.78		67.479		78.24	23	33	5.0	1.00			27.9
22	25.346	642.45	.00050457	1.945	10.27	514.2	8.287	16.12		85.114		62.05	21	32	5.0	1.00			31.0
23	22.572	506.49	.00040015	1.542	8.142	648.4	13.18	20.32		107.29		49.21	18	26	5.0	1.00			34.4
24	20.101	404.04	.00031733	1.223	6.457	817.6	20.95	25.63		135.33		39.02	16	20					38.2
25	17.901	320.42	.00025166	.9699	5.121	1031	32.97	32.31		170.59		30.95	12	16					42.0
26	15.940	254.10	.00019958	.7692	4.061	1300	52.97	40.75		215.16		24.54							47.0
27	14.196	201.52	.00015827	.6100	3.221	1639	84.23	51.38		271.29		19.46							52.0
28	12.642	159.81	.00012551	.4837	2.554	2067	133.9	64.79		342.09		15.43							57.0
29	11.258	126.74	.000099536	.3836	2.025	2607	213.0	81.70		431.37		12.24							63.4
30	10.025	100.51	.000078936	.3042	1.606	3287	338.6	103.0		543.84		9.707							70.1
31	8.928	79.70	.000062599	.2413	1.274	4145	538.4	129.9		685.87		7.698							77.1
32	7.950	63.20	.000049643	.1913	1.010	5227	856.2	163.8		864.87		6.105							84.6
33	7.080	50.13	.000039368	.1517	.801	6591	1361	206.6		1090.8		4.841							91.9
34	6.305	39.75	.000031221	.1203	.635	8311	2165	260.5		1375.5		3.839							101.6
35	5.615	31.52	.000024759	.09543	.504	10482	3441	328.4		1734.0		3.045							112.1
36	5.000	25.00	.000019635	.07568	.400	13217	5473	414.2		2187.0		2.414							119.7
37	4.453	19.83	.000015574	.06001	.317	16666	8702	522.2		2757.3		1.915							130.6
38	3.965	15.72	.000012345	.04759	.251	21015	13870	658.5		3476.8		1.519							140.6
39	3.531	12.46	.0000097923	.03774	.199	26500	22000	830.4		4384.5		1.204							151.0
40	3.145	9.88	.0000077634	.02993	.158	33416	34980	1047.0		5528.2		.955							163.4

Resistance in International Ohms at 68° F., according to Matthiessen's Standard of Resistivity.

Temperature Coefficients — Resistance per Mil-Foot, Ohms.

Temp. in Deg. C.	Resistance per Mil-Foot, Ohms
0	1.00000
1	1.00387
2	1.00776
3	1.01166
4	1.01558
5	1.01950
6	1.02343
7	1.02738
8	1.03134
9	1.03531
10	1.03929
11	1.04328
12	1.04728
13	1.05129
14	1.05532
15	1.05935
16	1.06339
17	1.06745
18	1.07152
19	1.07559
20	1.07968
21	1.08378
22	1.08788
23	1.09200
24	1.09612
25	1.10026
26	1.10440
27	1.10856
28	1.11272
29	1.11689
30	1.12107
40	1.16332
50	1.20625
60	1.24965
70	1.29327
80	1.33681
90	1.37995
100	1.42231

WIRE TABLE. INSULATED WIRE.

B.W.G.	B. & S.	inch	mm	SC Thickness of insulation. Inch.	SC Ratio of bare diameter to thickness of insulation.	SC Weight of insulation per 100 lbs. of covered wire.	SC Weight of covered wire per lb. of bare wire, k_s.	DC Thickness of insulation. Inch.	DC Ratio of bare diameter to thickness of insulation.	DC Weight of insulation per 100 lbs. of covered wire.	DC Weight of covered wire per lb. of bare wire, k_s.
1		.300	7.62	..	..	..	..	.020	15	2.28	1.022
	1	.289	7.34	..	..	..	..	.020	14.45	2.32	1.023
2		.284	7.21	..	..	..	..	.020	14.2	2.33	1 023
3		.259	6.58	..	..	..	..	.020	12.95	2.40	1.024
	2	.258	6.55	..	..	..	..	.020	12.9	2.40	1.024
4		.238	6.04	..	..	..	..	.020	11.9	2.50	1.025
	3	.229	5.82	..	..	..	..	.020	11.45	2.55	1.025
5		.220	5.59	..	..	..	..	.020	11	2.65	1.026
	4	.204	5.18	.012	17	2.20	1.022	020	10.2	2.85	1.028
6		.203	5.16	012	16.9	2.20	1.022	.020	10.15	2.86	1.028
	5	.182	4.62	.012	15.15	2.27	1 0227	.018	10.1	2.87	1.028
7		.180	4.57	.012	15	2.28	1.0228	.018	10	2.90	1.029
8		.165	4.19	.012	13.75	2.33	1 0233	.018	9 17	3.20	1.032
	6	.162	4.12	.010	16.2	2 24	1 0224	.018	9	3.25	1.032
9		.148	3.76	.010	14.8	2.30	1 023	.016	9.25	3.15	1.031
	7	.144	3.66	.010	14.4	2 32	1.0232	.016	9	3.25	1.032
10		.134	3.40	.010	13.4	2.36	1 0236	.016	8.4	3.55	1.035
	8	.1285	3.27	.010	12 85	2.40	1.024	.016	8	3.75	1.037
11		.120	3.05	.010	12	2.50	1.025	.016	7.5	4.10	1 041
	9	.1144	2 91	.010	11.4	2.55	1.0255	.016	7.1	4.35	1 043
12		.109	2.77	.010	10.9	2.66	1.0266	.016	6.8	4.60	1.046
	10	.102	2 59	.010	10.2	2 85	1.0285	.016	8.4	5.00	1.05
13		.095	2.41	.010	9.5	3.10	1.031	.016	5.9	5.55	1.055
	11	.091	2.31	.010	9.1	3.25	1.0325	.016	5.7	5.85	1.058
14		.083	2.11	.007	12	2.50	1.025	.016	5.2	6.60	1.066
	12	.081	2.06	.007	11.6	2.54	1.0254	.016	5.1	6.80	1.068
15	13	.072	1.83	.007	10.3	2.80	1.028	.016	4.5	7.80	1.078
16		.065	1.65	.007	9.3	3.15	1.0315	.016	4.1	8.60	1.086
	14	064	1.63	.007	9 1	3.25	1.0325	.016	4	8.80	1.088
17		.058	1.47	.007	8.3	3.60	1.036	.014	4 1	8.60	1.086
	15	.057	1.45	007	8.1	3.70	1.037	.014	4 1	8.60	1.086
	16	.051	1.30	.007	7.3	4.20	1.042	.014	3.6	9 60	1.096
18		.049	1.25	.007	7	4.40	1 044	.014	3.5	9.80	1.098
	17	.045	1.15	.005	9	3.25	1.0325	.012	3.75	9.30	1.093
19		.042	1.07	.005	8.4	3.55	1.0355	.012	3 5	9.80	1 098
	18	.040	1.02	.005	8	3.75	1.0375	.012	3 33	10.10	1 101
	19	036	0.91	.005	7.2	4.30	1.043	.005*	7.2	5.60	1.056
20		035	0 89	.005	7	4.40	1.044	.005*	7	6.00	1.06
21	20	032	0.81	.005	6.4	5.00	1.05	.005*	6.4	6.60	1.066
22	21	028	0 71	.005	5.6	6.00	1 06	.004*	7	6.00	1 06
23	22	.025	0 64	.005	5	7.00	1.07	.004*	6.25	7.00	1 07
24	23	.022	0.56	.005	4.4	8.00	1.08	.004*	5.5	8.00	1 08
25	24	.020	0.51	.005	4	8.80	1.088	.004*	5	8.80	1.088
26	25	.018	0.46	.005	3.6	9.60	1.096	.004*	4.5	9.60	1.096
27	26	016	0.41	.005	3.2	10.40	1.104	.004*	4	10.40	1.104
28	27	.014	0.36	.005	2.8	11.25	1.1125	.004*	3.5	11.25	1.112
29	28	.013	0.33	.005	2 6	11.65	1.1165	.004*	3.25	11.65	1.116
30		.012	0.31	.005	2.4	12.05	1.1205	.004*	3	12.05	1.120
	29	.011	0.28	.005	2.2	12.45	1.1245	.004*	2.75	12.45	1.124

* Double silk : 1 mil of silk insulation taken equal in weight to 1.25 mil of cotton covering.